PRO PAINT & BODY

Jim Richardson

with **Tom Horvath**

HPBooks

HPBooks

Published by the Penguin Group

Penguin Group (USA) Inc.

375 Hudson Street, New York, New York 10014, USA

Penguin Group (Canada), 90 Eglinton Avenue East, Suite 700, Toronto, Ontario M4P 2Y3, Canada
(a division of Pearson Penguin Canada Inc.)
Penguin Books Ltd., 80 Strand, London WC2R 0RL, England
Penguin Group Ireland, 25 St. Stephen's Green, Dublin 2, Ireland (a division of Penguin Books Ltd.)
Penguin Group (Australia), 250 Camberwell Road, Camberwell, Victoria 3124, Australia
(a division of Pearson Australia Group Pty. Ltd.)
Penguin Books India Pvt. Ltd., 11 Community Centre, Panchsheel Park, New Delhi—110 017, India
Penguin Group (NZ), 67 Apollo Drive, Rosedale, North Shore 0632, New Zealand
(a division of Pearson New Zealand Ltd.)
Penguin Books (South Africa) (Pty.) Ltd., 24 Sturdee Avenue, Rosebank, Johannesburg 2196, South Africa

Penguin Books Ltd., Registered Offices: 80 Strand, London WC2R 0RL, England

While the author has made every effort to provide accurate telephone numbers and Internet addresses at the time of publication, neither the publisher nor the author assumes any responsibility for errors, or for changes that occur after publication. Further, the publisher does not have any control over and does not assume any responsibility for author or third-party websites or their content.

PRO PAINT & BODY

PRINTING HISTORY
First HPBooks edition / September 2002
Revised HPBooks edition / February 2011

ISBN: 978-1-55788-563-0

PRINTED IN THE UNITED STATES OF AMERICA

10 9 8 7 6 5 4 3 2 1

NOTICE: The information in this book is true and complete to the best of our knowledge. All recommendations on parts and procedures are made without any guarantees on the part of the author or the publisher. Tampering with, altering, modifying, or removing any emissions-control device is a violation of federal law. Author and publisher disclaim all liability incurred in connection with the use of this information. We recognize that some words, engine names, model names, and designations mentioned in this book are the property of the trademark holder and are used for identification purposes only. This is not an official publication.

MAY 2011

CONTENTS

ACKNOWLEDGMENTS

Top body and paint pros Tom Horvath, Junior Conway and Bruce Haye made this book—and the new chapters on the latest techniques and technologies—possible. I owe them a great deal of thanks for sharing their knowledge and experience so others can learn the methods and tricks that have taken them years to develop and made them the top people in their profession. I would also like to thank my wife, Bette, for keeping our lives going while I was spending long hours researching and writing this book. And finally, I want to dedicate this book to my father, who, among other things, was a pretty fair panel beater himself.

This book is for those who want to sit at the feet of the masters and learn how they do automotive finish work worthy of Pebble Beach and the Oakland Roadster show. It's a chance to acquire the techniques, tips and tricks talented artists have developed over the years. I'm talking about techniques that can make the difference between creating a car that looks good and creating one that looks fabulous.

This book is also for those who want to learn about the latest automotive paint technology. In the last twenty years, panel beating and especially painting have changed dramatically as a result of new environmental laws, and better, newer materials. I'll show you how the new paints and primers work and how to use them to get the kind of finish you are looking for.

Find out what great master craftsmen like Junior Conway and Tom Horvath do to make their show-winning creations. Just keep in mind that learning their secrets is only part of the process because quality paint and finish are a combination of arts that take practice and patience to master. To do all of it right you need to know how to weld, work metal, prime, paint, polish and detail a car, and that requires mastery of a number of skills.

None of these skills is hard to acquire in itself, but when added together they amount to a fair accumulation of knowledge and experience. And of course, you need not master all of them. Welding can be farmed out, as can metal finishing. And there are those who do nothing but shoot paint beautifully. Any of these skills is very marketable as long as you stay current.

You need to know how the new polyester primers work, how urethane paint is applied, how waterborne paints are used and how different kinds of metals respond to being worked. And you need to know how to ply your trade or passion safely, because you must deal with some rather toxic materials when cosmetically restoring cars.

So if you too want to acquire show-winning techniques, first read the safety chapter, then read for example, a chapter on welding or panel beating and go out and try it. Get a feel for metal and what you can and can't do with it. Body shops all over town have bins of dented and crushed panels they will gladly give you to practice on.

And don't be afraid to ask questions. Even the legendary Junior Conway—whose cars have graced the covers of rod and custom car magazines for years (as well as the back cover of this book) took the time to give me some tips. And why not? His reputation as the best automotive painter in Southern California—arguably the United States—is unassailable, and he knows that very few will ever master the art of painting cars as well as he has, even if they know his secrets.

My coauthor for this book is Tom Horvath, who for many years owned Tom's Custom Auto Body in Anaheim, California, before becoming a pioneer in the CSI Clearcoat Solutions polishing system (see Chapter 32). He has spent countless hours teaching me his show-winning body and paint techniques. He is truly a master and a good teacher. And his magnificently finished Ferraris, Rolls Royces, Panteras and Porsches have won rooms full of trophies including some from Pebble Beach.

Tom has perfected color sanding and polishing techniques and products that he has made available to other paint pros and the general public, and he tells us how he uses them to get those flawless finishes his work always exhibits. We also tell you about other products that Tom and I have found to be useful and time saving, and let you know where you can get them.

As for my own painting skills, I grew up around panel beating and have it in my blood—probably literally—since real men didn't wear respirators in the old days, and my dad once shot paint for Howard "Dutch" Darrin. Darren was doing his one-off Packard roadsters back in the 1940s. Dad also painted cars in his own garage for extra money.

When I was a little kid, I remember Pop bringing home

rusting, derelict cars, and working out their dents, patching their panels and painting them so they looked new again. I also remember him humping over their front fenders and making them mechanically road worthy, though that was not his forte. He was truly a master painter though, and could actually shoot on a lacquer finish so it needed almost no polishing, and then rub it out to make it dazzling.

As I grew up I picked up some of the old man's tricks and have restored a number of cars and taken home some gold myself, though I never worked as a professional. But then I realized I needed to update my skills too. And with the help of pros like Tom Horvath, Junior Conway and my buddy Bruce Haye down in New Zealand, I have discovered that the work is now both easier and more difficult, depending on which part of the process you are talking about.

Since the first edition was published in 2002, there have

been new plastic fillers, polyester primers and urethane paints developed that make life easier for the auto finishing professional, but they require careful techniques, and their chemical makeup means they must be mixed and applied exactly according to the instructions if you want good results. This revised and updated edition includes new chapters (11, 16, 19, 23, 24, and 26) as well as new photos and information throughout.

In the area of bodywork, welding has become somewhat easier thanks to MIG and TIG technology, but it requires learning new techniques as well.

That being said, if you can drive a nail without bending it, you can do bodywork. If you can follow instructions and are willing to practice, you can paint. And if you are willing to put in the time it takes to rub out and polish a car, you too can achieve show-winning results. Read on and we'll tell you how.—*Jim Richardson*

Chapter 1

Safety First

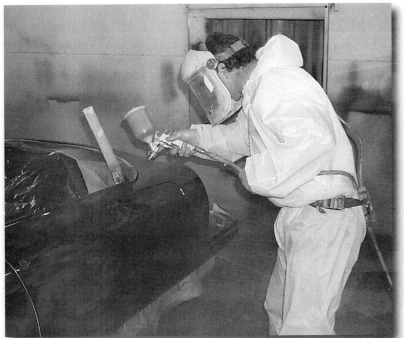

This painter wears a full plastic suit, gloves and a positive pressure mask while spraying urethanes.

Painting cars can be hazardous! Therefore, you must MAKE SAFETY A HABIT. Just about every high school shop area in this country has a sign hanging in a prominent place with these words on it. We have heard and seen this phrase so often that many of us have stopped paying any attention to it, and that is dangerous. Especially when doing bodywork and painting. Notice that the sign doesn't just say "Be careful." Instead it instructs us to do things the safest way every time until doing them that way becomes a habit. Just being careful isn't enough, because that can divide your attention.

Hear Me Out!

Bodywork, welding and painting are hazardous to your health if you don't follow the correct procedures. When grinding and sanding on car panels, always wear at least a particle mask because the stuff you are grinding off will be spun into the air and will most likely be toxic. And even if it isn't, you don't want to fill your lungs with it. Wear hearing protection too. The insidious thing about excess noise is that it will damage your hearing without you ever realizing it. The process is painless, and until appreciable damage is done, you'll never know anything has happened to you.

Not Seeing Is Believing

Eye protection should be obvious for any grinding, cutting or welding work. One hot shard from a die grinder can blind you. It is especially critical that you wear the right protection

This is the Survivair pump, which provides an adequate supply of fresh clean air to as many as two painters at a time.

while welding, and arc, MIG or TIG welding requires a darker lens than gas welding, so don't mix them up and use the wrong one. In fact, if you are going to be doing much arc welding, you need a complete facemask if you don't want to wind up with a major sunburn at the end of the day.

You can pull your breathing air supply from your compressor, but you will need special filters to take out the oil and moisture if you do.

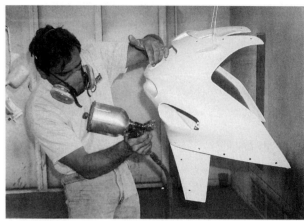

Never paint without at least a respirator equipped with fresh charcoal cartridges.

This is Survivair's full face mask with positive pressure air from the outside. These are an absolute must if you wear a beard.

Where There's Smoke...

Solvents, paints and welding equipment can combine to create an explosive situation. It's a little like smoking a cigar on the Hindenburg. Make sure your shop has plenty of ventilation. A large fan with an explosion-proof motor and connector open to the outdoors is a must if you cannot actually work outdoors or in a garage with the door open. Fire extinguishers are an obvious necessity too. Post several heavy-duty types close at hand around your work area. Make all electrical connections for power equipment, switches and lighting are clean, tight and well insulated.

If you are working at home, move your hot water heater outdoors if it is in your garage because a pilot light could easily start a fire if fumes build up. Keep paints and solvents in metal containers away from potential ignition sources. I use an old refrigerator that no longer works as a locker to store paints and solvents, because the thing is all metal on the outside, has good insulation against heat built right into it and has a tight rubber seal to contain fumes.

Give Me Air

All automotive paint is toxic. Even the old lacquer solvents and thinners caused brain damage, and the new waterborne paint contains acetone among other nasty things. As we've said before, the new urethanes are particularly dangerous because the hardeners used with them contain isocyanates, which are cyanide. The stuff is tasteless and odorless, and if you breathe it you run the risk of kidney failure and rapid death. There are also heavy metals, etching acids and other toxins in paint that are almost as insidious and they are invisible. Much of this stuff is readily absorbed through your skin and mucous membranes too, so just wearing a respirator won't completely protect you.

To do the job safely you need a nylon painting suit, plastic gloves (not latex, which will react with some solvents) and a respirator hooked up to an outside air source. This air can come from your compressor through filters to eliminate moisture and oil contamination, or better yet, you can use a separate outside air source like those sold by Survivair in California that gives you a steady source of clean filtered air at all times.

Either way, the piped-in outside air creates a constant positive atmosphere that keeps deadly fumes out. This is much better than an ordinary respirator because as you inhale with a typical respirator you create a negative atmosphere that will pull fumes in if your mask isn't sealed perfectly around your face. Also, the typical respirator doesn't protect your eyes.

Survivair also sells top quality full face masks that protect your eyes, and if the mask seals properly around your face, your respiratory system will be protected from most paint problems—that is, provided you change the filter canisters regularly, as specified by the manufacturer.

If you do a lot of painting, get a gun cleaning tank. But never submerge a spray gun in solvent.

Some old-time professional painters will tell you that—when you have your respirator in place—if you smell paint, the filters have clogged and need to be changed. But that's too late to completely protect you. If you smell paint, you are breathing poison. Fact is, if you smell fumes while spraying paint, something is wrong and you are not being protected adequately.

Remember, you can't smell the truly dangerous ingredients of paint. Stop spraying and fix the problem, whether it is a faulty seal or clogged filters. Don't try to finish out the paint in your gun cup at that point either because you can have problems even with brief encounters with isocyanates.

If you wear a beard as I do, you definitely need a Survivair outside air system to produce a positive atmosphere in your mask because a respirator won't seal completely over hair. When you inhale you pull in paint fumes because of the faulty seal. If you can't afford a Survivair system, shave, and smear a little Vaseline around the sealing lip of your respirator in any case. Some toxins such as heavy metals build up in your system over time, so you won't even remember what hit you when you become ill.

If you wash a lot of greasy parts, use a metal parts washing tank with a proper hood. Never smoke around the parts tank and keep it closed when you are not using it. Once again, make sure the parts washer is kept well away from spark or flame. I know of a fellow who actually decided to weld on top of a parts washer and attached the ground for the welder to the lid of the tank. I won't make any remarks about his lack of common sense because it is disrespectful to criticize the departed, but if you don't see the problem with this, have someone else work on your car.

Keep It Clean

Oily rags, open containers of flammable liquid, and cords and hoses running all over the floor of your shop are accidents waiting to happen. Ask any paramedic about how many people he or she has taken to the emergency room burned, scraped, bashed or with something in their eyes. It's good for all of us to hear a few horror stories now and then. But as we said at the outset, the most important strategy to avoid injury is to MAKE SAFETY A HABIT.

Chapter 2
Evaluating a Project Car

The inspection of a project car should start from the bottom. People rarely try to disguise rust and sloppy repairs from underneath. Rotted frames are bad news. So are rusted-out floors.

Not too long ago I was at a friend's restoration shop when a fellow pulled up in a '60 Cadillac hearse that had been made into a pickup truck. It was dented, sagging, dirty and tired. I doubt if it would even have made a good parts car. Its proud owner stepped in the door of the office and inquired as to how much it would cost to have the shop restore it.

At first my friend thought the fellow was joking, but after having a good laugh we noticed an uncomfortable look on the customer's face that indicated he wasn't being funny. He told us that he just loved the old Cad, and that it had belonged to his father. My friend tossed out a ballpark figure in the neighborhood of $40,000. That number nearly knocked the guy off his feet, and the fellow beat a hasty, mumbling retreat.

This little tale illustrates a couple of things: the first being that almost any car can be restored with enough work and money. The second lesson—one you probably already know—is that one man's junk may be another man's treasure. If a particular car means more to you than any other, and you don't care if you ever recoup even a portion of your costs and effort, you can make even a derelict such as that old hearse new again. The point of this chapter isn't to tell you which makes of cars are good or bad, or even what can be fixed and what cannot. It is here to help you go into a project car with your eyes open, so you won't wind up with less—or more—than you bargained for.

Keep in mind that, unless you are working on a one-off Ferrari or some other rare exotic, restoring it mechanically

Rusted subframes on unit-bodied cars are especially difficult to deal with and can actually affect the structural integrity of the vehicle.

won't be nearly as big a task as restoring it cosmetically. Certainly, before you buy a prospective project you will want to know all about its mechanical condition, but more important—and much more costly to deal with—will be the body, paint and cosmetic aspects of your project.

Rust is enemy number one. Properly repairing a badly

If we hadn't opened the trunk we'd have never known how badly damaged its lid was. Don't forget to pull mats back and check trunk floor for rust too.

This is an especially obvious case of eyebrow rust, common on cars from the '50s and '60s. Moisture gets trapped in these areas during driving and eats away at upper fenders.

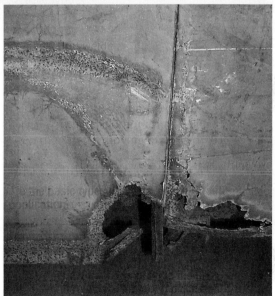

I believe that the only pieces worth saving on this car are those two stainless strips along the belt line. Of course, if the car is a classic convertible it might be worth an attempt, but it would be a daunting task.

rusted car taxes the skills of even the most experienced and gifted panel beater. This is especially true for unit-bodied cars because as you cut panels away, you also cut away structural integrity. Unless you keep everything in alignment, using custom fabricated braces and jigs, the car's body will warp completely out of shape. The only time it might be worthwhile to fix an extensively rusted unit-bodied car is if it is a rare and sought-after classic of some sort.

Otherwise you would be far better off waiting and buying a relatively rust-free car, even if you have to pay more for it. Fabricating panels, welding them in and making sure everything is right can be very expensive. And that's especially the case if you have to pay to have the work done because the job requires a lot of skill, and the time to do it right. On the other hand, replacing a small section of rusty floor pan or a couple of rotting lower kick panels isn't a big deal if you know what you are doing.

Start at the Bottom

Before you start checking body tin, take a good look underneath the car. Place a tarp or ground sheet you don't mind getting dirty beside the car and have a look at the chassis or subframe using your flashlight. Look for rust in floor pans, structural members and fender wells. Be sure to look for previous rust repairs also, using your refrigerator magnet.

It is surprising what some people have done to old cars to get a few more miles out of them. I have actually seen whole floor pans pop riveted in, then covered with undercoating to disguise the repair and prevent further rust. Look for poor welds too. The fact that a panel or structural member has been repaired is not a problem in itself if the repair was

done properly, but such repairs do tell you that the car has spent a fair amount of time out in wet or snowy weather, or in a damp climate.

I have even seen cars where serious rust holes have been stuffed with plastic filler (Bondo) and painted. Plastic filler has an important place in body and fender repair, but it has no structural integrity at all, and will eventually crack and fall out if stressed or flexed. Structural rust is sometimes referred to as body cancer and it is an apt analogy because if you don't get rid of it completely it will continue to grow and eat away at your car's precious metal.

Don't let the fact that you are looking at a "California" or "Southern" car fool you into complacency either. I live in California and I have seen some pretty nasty rust damage on cars that have never left the state. Where a car originates from has less to do with its condition than how it has been stored and how well it has been cared for. It pays to go over any prospective purchase—even the pristine-looking ones—carefully before you lay down your money.

Don't forget to check station wagons and hatchbacks for rust around rear hatch. Rubber seals often rot and let in moisture.

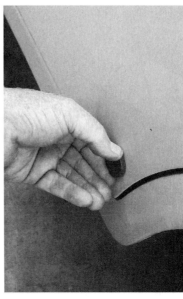

A small household magnet is ideal for checking lower kick panels, doors and rear quarter panels for plastic filler. If the magnet sticks, chances are the panel is healthy.

Open the Hood and Trunk

Another important place to check for rust is around the shock towers in the engine room and in the trunk. If you see damage in these places, avoid the car unless you are prepared to do some very extensive and expensive fabrication work. Also, look for amateur welds in these areas indicating previous repairs. Such repairs can be fine if the owner did the job right and then took the time to rust-proof the welds and carefully coat the repair. But if he didn't, the welded-in patch will rust out just that much faster due to the oxidizing heat created by the welding process.

Check the Panels

Start checking the body by looking for obvious rust problems. Look around headlight eyebrows, kick panels, rocker panels, drip moldings and under trunk lids. Each make and body style has its peculiarities and vulnerabilities. Look carefully anywhere you see creases or places where water can collect. Open each door and check along the bottom for rust-through. Many older cars are designed to let a certain amount of water run down into them and out drain holes at the bottom. If the drain holes get plugged with dirt though, the doors rust out.

It isn't enough just to feel along these door bottoms. Get down and inspect them directly or use a small mirror to have a closer look. Stopgap repairs could have been made with plastic filler. If the sills are good, check the front hinge pockets for rust, wear and damage. Gently pull up on each door to determine how much play is in the hinges and mountings. A little bit of hinge sag can be corrected with new hinge pins but it indicates that the car has had a lot of use. Close each door and check whether it engages the striker plate properly or not.

Roll each window up and down to see if it operates smoothly. Inspect whisker moldings for wear, and make sure the window is held securely in its tracks. Look at all the glass for the correct sandblasted-in bugs or logos. An odd bug or no logo at all means that the window has been replaced. That's no problem in itself, but it is worth noting because the replacement may have been part of a much bigger repair due to collision.

Look carefully around the base of the windshield and back window for rust and marginal repairs. Open the hood and check the fender wells and pans. Also check the hood underneath for fire damage and sagging sound deadener. A carb fire can harden a hood and cause major rust and warping problems. Check to see that the hood hinges work correctly and hold the hood up if they were designed to do that. Close the hood and check for fit all the way around. If the hood is on cockeyed it may be that it was removed, then installed out of position, or it could mean the car has been hit in front and a fender is knocked out of place.

Open the trunk and check for rust and repairs. Pull back the matting and check for rust-out in the floor. Look at the recessed molding for the trunk seal. Is it rusty and damaged? Is the seal hard and cracked or torn? The seal is easy to fix, but a rusted-out molding is not so easy. The trunk is also a good place to check for original color if the car has been repainted too.

Magnetic Checks

Badly done amateur repairs using filler can often be spotted by letting your eye follow the highlights down the car, but the easiest and best way to check for rust is with a little refrigerator magnet. Don't use a more powerful one because it can actually stick to metal that is 1/8" or more below the paint

This bubbling mess is a plastic filler repair that has not been properly moisture-proofed. Our household magnet won't stick.

Poorly aligned panels indicate possible collision damage and sloppy repair work.

A Hawkeye Borescope is ideal for finding rust in tight places. Just look through the eyepiece.

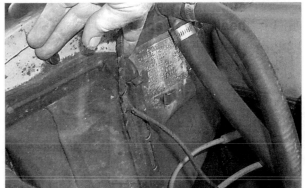

You can tell what color a car was originally, as well as how it was equipped, by looking at the cowl tab and the build plate, then comparing these to reference manuals.

surface. Check anywhere you see a ripple or irregularity.

If you see no obvious problems, get down low and sight along body panels for funny bumps. Then, even if the body looks straight and clean, run your magnet along kick panels, lower door areas and rocker panels. If the magnet refuses to stick anywhere, it tells you that a repair has been done using plastic.

Unroll your ball of string and have a friend hold one end while you pull it tight. Now hold it parallel to the car about four inches away from the body and look for any alignment problems. If the car is a straight, slab-sided type, you can easily spot tweaked front fenders. Next, hold it down low and check to see that body panels line up along the bottom. If panels don't line up it may be because they were removed, then installed incorrectly, or that the structure under the panel is bent due to collision.

Check Alignment

Check panel gaps around doors, hood, trunk and fenders. If you see gaps that are too tight or too loose, again, they may have been repaired and then installed incorrectly, or the car may have been in a collision that knocked its frame or subframe out of alignment. You can even get down and measure the length of the car from front to rear using the string to see if there is a difference from one side of the car to the other, indicating an alignment problem or possible bent frame.

Hawkeye

If you are really serious about checking a car out, you can use a Hawkeye Borescope to peer into inaccessible areas for rust and problems. These devices are made for checking rifle bores and use fiber optics to send light through a tiny lens at the end of a probe that can be focused. The tool isn't cheap, but it could save you money in the long run if you are going to be working on cars regularly because it can also be used to check inside combustion chambers in engines too, by inserting the probe into the spark plug holes.

Paint Problems

On older cars it is actually better if they have an oxidized and dull original finish than if they have a fresh cheap paint job on them that looks good. That's

Even though it had to be torn down completely and the rocker panels cut out, this E-model Jag will be worth the effort when it is finished because it is a rare, high-dollar car.

CUTTING COSTS

Obviously, the more work you can do yourself the less expensive your paint job will be. The job still won't be free though, because you will need to pay for paint that could cost $300 or more per gallon, plus the thinners and primers to go with the paint, as well as sandpaper, stripper, body filler and the rest. But the materials are comparatively inexpensive when compared with labor costs at $60 to $90 dollars an hour.

Most of the work involved in making a car beautiful isn't that difficult to master. It just takes time and patience. But if you aren't prepared to practice certain skills, or don't feel confident about some area of the project, you will need to factor in the costs of sending the work out. The first order of business is to take inventory of your skills and consider your willingness to learn. Even if you do no more than take off the brightwork and strip the car of old paint you can save a fair amount of money.

Welding—If you can't weld, consider taking a class at the local junior college. Good welding is the key to rust repair, and you will need to know how to wield a torch to shrink panels and do lead work. If you aren't prepared to do a little seat time practicing with a MIG welder and acetylene torch, you will want to farm this kind of work out because you can ruin panels if you don't know what you are doing.

Panel Beating—Again, you will need to spend a little time developing skills in order to do this kind of work properly. Unless your standards are very low, you can't just bash dents into submission and slather on the plastic filler if you want a decent job. If you aren't willing to learn a few tricks and strategies before you start knocking on tin, you'd be better off having someone else do this work.

Painting—Shooting paint is a trade all by itself. The best painters can shoot on a beautiful finish that needs no rubbing—unless you are looking for jewel-like show-winning perfection. The rest of us can shoot paint, block sand and rub the job out and still arrive at a beautiful paint job but it takes us a little longer.

Polishing—With a little care and a lot of effort anyone can make their car look beautiful with color sanding and polishing. Here is another area where the comparative novice with limited means and skill can save some money and make his car look gorgeous.

Planning—Once you have assessed your present skill level and what you would be willing to learn, get some estimates from the people to whom you will be farming out work to determine what they might charge to do their part. In the end, if a person of modest means is willing to put in the time and take on the adventure and challenge of learning a few tricks, they can have a show-winning paint job that won't empty their savings account. But to avoid problems you need to have a clear picture of the situation going in.

because you can often scuff and paint over an original finish pretty easily, but you won't want to do that to a cheap second paint job, or even an expensive one, for a couple of reasons. The biggest problem with repainted cars is that the paint that is on the panel is generally too thick to take another coat.

Another problem with painting over old work is that you don't know how well the car was prepared before the previous paint was applied. Anything you put on top of a cheap paint job will only adhere as well as the paint underneath. There are also occasional problems with compatibility between the paints of the past and the new ones being used today. And finally, if the paint on the car is more than 4 mils thick it will be prone to cracking and flaking in any case.

Tom Horvath sets up a car on a frame rack to check it out.

Chapter 3
Frame Alignment

Everyone has seen a car at some time or other going down the road with its back end hanging out to one side or the other, even though the car's front end is proceeding in a straight line. This phenomenon is called *dog-tracking*. It usually happens because a collision has bent the frame or body of the car, or because the front suspension or rear axle has been knocked cockeyed. The repercussions of the problem go well beyond just looking funny as you drive along though.

Cars that are out of alignment are dangerous to drive and can be almost uncontrollable if the misalignment is serious. Their tires wear out quickly too. Fixing the problem can be as simple as realigning the back axle on a car with a full frame, to pulling the whole car back into shape in the case of a unit-bodied car that has absorbed a bad hit.

Realigning a seriously damaged car is a task best left to the pros who have been trained on a frame rack. Heat and tremendous force is sometimes necessary, and that can lead to a dangerous situation. That being said, here is how to check out a car to determine what needs to be done. In some cases, after careful analysis you may discover that the task is more trouble than the car is worth. But if it is a high-dollar classic you will probably want to do whatever it takes to bring it back.

Check It Out

Most cars have reference points such as holes or dimples in their frames and bodies from which you can measure to

determine their overall alignment. You can do a quick, rough check of a car's alignment using a measuring tape to check diagonally from fixed points across a car's frame or subframe. If, after a few rough calculations you suspect misalignment, you'll need to check things out further, and how you do that depends on whether your car has a full frame or is unit-bodied.

Cars with Frames

Full-framed cars with bent frames are easier to deal with and generally less critical than misaligned unit-bodied cars. A bent frame can cause a car's steering to be messed up, and can cause its body panels to fit together badly, but unless the damage is severe, the car's structural integrity will not usually be compromised. Here's how to check out and fix a bent frame:

Place the car on a level floor, and jack it up so its wheels are completely off the ground and so that no part of the car is supported by them. Put sturdy jack stands under the frame's center section and let the car down onto them. To measure accurately you'll need a tape measure and a tram gauge. You can make your own tram gauge out of a couple of three-foot-long pieces of 1" by 2" if you don't have a professional one. Cut slots in one end of each piece of wood and install screws with wing nuts. Then install a couple of long bolts with pointed tips at the other end of each piece.

To use this tool, extend the strips out diagonally across the frame so the bolts center in the index holes or points on the

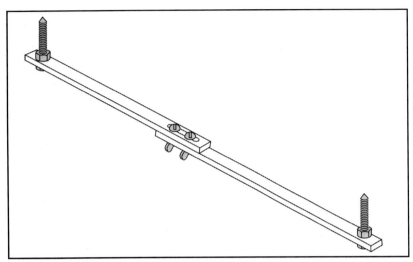

Pros use a tram gauge and tape measure to check the dimensions of frames. You can make a tram gauge yourself out of 1" X 2" wood strips and some common hardware.

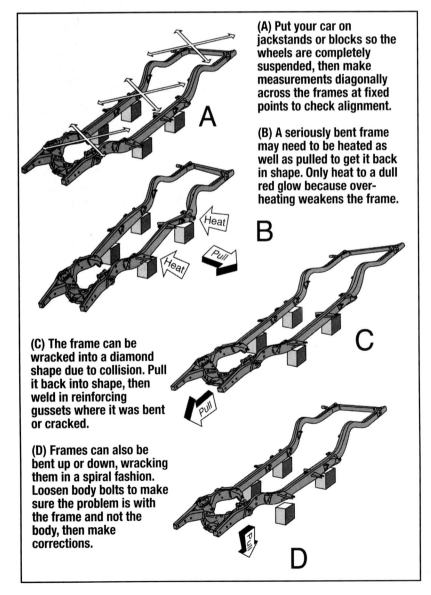

(A) Put your car on jackstands or blocks so the wheels are completely suspended, then make measurements diagonally across the frames at fixed points to check alignment.

(B) A seriously bent frame may need to be heated as well as pulled to get it back in shape. Only heat to a dull red glow because over-heating weakens the frame.

(C) The frame can be wracked into a diamond shape due to collision. Pull it back into shape, then weld in reinforcing gussets where it was bent or cracked.

(D) Frames can also be bent up or down, wracking them in a spiral fashion. Loosen body bolts to make sure the problem is with the frame and not the body, then make corrections.

frame. Mark the tool at its center with a pencil, then make a cross check diagonally to your original measurement. If you have to adjust the sticks more than 1/8 inch in or out, the frame is out of alignment.

Of course, a frame's rails can be parallel but one of them can be warped up or down. If the frame does not sit level on the jack stands, loosen the bolts holding the body to the frame. If the frame then sits flat on the stands, the body is warped. If the frame still does not sit flat after you loosen the body attach bolts, the frame is warped.

A frame can also be bent to a banana shape as a result of a collision from the side. To check for this, do the diagonal measurements, then measure across the rails. Use a carpenter's square to make sure your tape is at 90 degrees to the frame rails, and check at several points along the frame.

A car's frame can also become diamond-shaped due to a blow to one frame rail at the front or rear of the car. Measure diagonally across the center section of the frame, then measure its rear and front sections the same way. Sometimes all that is wrong is that one horn of the frame has been bent. Other times the entire frame is out of plumb.

People have realigned bent frames with chalk marks on the floor using just a come-along lashed around a stout pole, but this can be dangerous. If you don't have the use of a frame alignment rack, you would be wise to take your car to a professional, trained in frame alignment.

Sometimes frames can be tweaked back into shape cold just by doing a pull with the alignment device, but other times it is necessary to heat the frame in a couple of areas using an oxyacetylene torch. If you need to heat the frame to straighten it, consider taking the body off the car. If the frame is so badly out of alignment that heat is required, the body may well need work too. And the whole process of straightening the frame will be easier and safer with the body removed from the chassis.

Removing a car's body is a lot of work, but sometimes it's unavoidable. Begin the process by removing the front clip, consisting of fenders, grille and hood. Scribe around the hood hinges so you will know how to position it when you put it back in place, then get a couple of strong friends to hold the hood while you loosen it. Save any shims you find under the hinges and don't mix them up.

Remove the trunk lid the same way. Remove the doors at their hinges. Now remove the front and rear seats. Disconnect the speedometer from behind and remove its cable. Unplug the wire harness going to the front of the car. Disconnect any linkage to the clutch and transmission. Once you

1. First Tom zeroes out the rack so it is perfectly straight before any measurements are made.

2. Locating pins on the frame rack are carefully positioned in the index holes on the cowl of the car.

3. Pins underneath are then located in the proper holes in the car's subframe too.

4. Measurements are checked carefully and noted at each fixed point on the car.

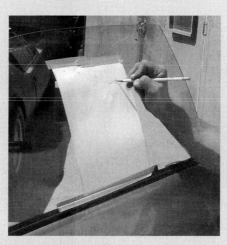

5. Tom taped sheets of paper to the side windows and noted down the dimensions for each side of the car.

6. Large clamps grip the pinch moldings of the body so it can be pulled back to its original dimension.

continued on next page

are sure there is nothing that will hang up when you lift the body, you are ready to lift it off.

If the body is not rigid enough to come off as a unit without bending or deforming, you will need to make X braces using steel angle iron and bolt them into place to stiffen it and hold it in alignment. Smaller, lighter car bodies can be lifted off by two to four strong lads, but larger car bodies will need to be jacked up at several points until the chassis can be rolled out from under it.

Follow the diagrams nearby to pull out various frame problems. When using a torch to heat the frame, only heat the metal to a deep red, not orange or white. The reason for this is that the frame will lose too much of its strength if you get it too hot.

Also, it is a good idea to weld in fillets or gussets to reinforce the areas where you heated, and to fix any cracks that might have occurred as a result of collision or pulling the frame back into alignment.

If you need to cold-bend a frame horn back into place or make some small adjustments to frame alignment with the body on the car, you will need to pull the frame just a bit further than where you want it to be, because steel has a memory and will spring back to a degree. This phenomenon is hard to judge. The process takes experience to get it right. Make moderate pulls first, and check your measurements frequently. Gradually increase the pull until you get the result you want.

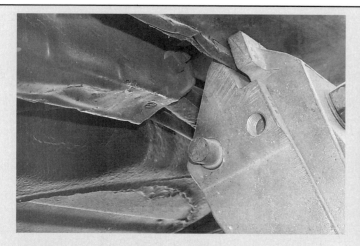

7. Clamps hold the car firmly because considerable force may need to be applied.

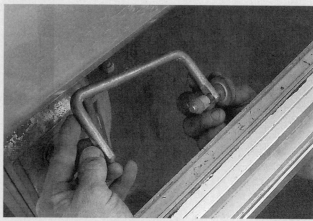

8. These clamps hold the frame rack in proper alignment while a pull is being made.

9. Tapping in critical areas on the car's subframe sets up a harmonic that relieves stresses and the car comes right.

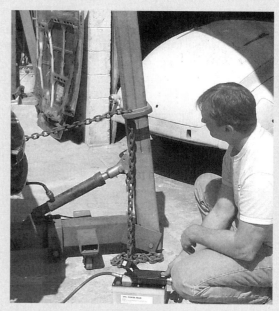

10. Tom takes up the tension, then makes a modest pull. Nothing much happens.

Unit-Bodied Cars

Collision damage to unit-bodied cars is a much more serious problem than it is to a car with a heavy steel frame. That's because a unit-bodied car is sort of like a cardboard box in that it depends on its shape to maintain its strength and rigidity. Crush one corner of a cardboard box and you seriously compromise the whole box's structural integrity. And yes, you can push the box back into shape, but the damage is done. The box will never again hold its shape under stress the way it once did.

Unit-bodied cars depend on their metal skins and their subframes—also made of thin, high carbon sheet metal—for their strength. If a unit-bodied car is smashed or badly damaged it loses its rigidity and becomes dangerous to drive. Worse than that is the fact that even though you can pull the car back into its original shape, it may still be a danger to itself and others because its panels will be work-hardened, bent and probably torn.

Also, unit-bodied cars need to be aligned much more precisely because very little can be done by adjusting the suspension to correct for errors. Unit-bodied cars are usually repairable though, if the damage isn't too severe, but to do it you need a frame rack especially designed to do the job. Tom Horvath's son got slammed by a Suburban while out cruising in his Ford Probe. Remarkably nobody got hurt, but the Probe was wracked completely out of shape.

The first thing to do when straightening a car is to determine what caused the damage. Just as with taking out a dent, you need to know from which direction the car was hit and how that force was transferred through the car. In many cases with unit-bodied cars, the whole body is wracked out of alignment after a serious collision. Diagramming the impact in your mind can take you a long way toward figuring out what needs to be done.

Tom Horvath's frame rack was set up to take measurements from fixed reference points on the car, but other types of frame racks use optical sighting mechanisms and spirit levels. Tom's rack was also set up to grip and pull a unit-bodied car from the pinch moldings underneath, where the car is welded together. A bar under the car with holes in it is used in conjunction with another arm that has pins that can be located in the holes of the arm to make the hydraulic jack pull at the correct angle.

Tom started by taping a piece of paper on each of the car's side windows so he could write down each of his measurements. He then verified that the frame rack itself was square with the use of a carpenter's angle square. Next, he started establishing the pins on the ends of the measuring arms of the rack in the reference holes in the car's body. He then took measurements at the front end, at the cowl and at the rear of the car and noted them down.

What Tom discovered was that the car had been hit in such a way as to knock the whole body into a slightly trapezoid shape when viewed from the front. The car had also been wrapped into a slight banana shape when viewed from the top. Tom surmised that pulling the car straight along its length first might also correct the trapezoid effect in the bargain. He then hooked up the hydraulic puller on the frame rack, took up the tension and made a modest pull. Nothing much happened.

Tom then took a large hammer and tapped in key places along the car's structure and subframe. The metal started to straighten and pop as it was pulled out straight. Getting the car back in alignment would be an involved process requiring patience and professional acumen, but it would be realigned properly. Tapping with a hammer sets up a harmonic in the metal that helps release the tension and stress that has been generated by the collision.

Once the car is square and in alignment again the real fun begins. Gussets and reinforcing patches will need to be welded in to strengthen damaged areas, and panels that have been warped and bent to the point where they no longer provide the necessary strength will have to be cut out and new pieces welded in. The doors, hood and trunk lid will have to be re-fitted and any broken glass will have to be replaced. After that, all of the welds will have to be cleaned, derusted and painted with epoxy primer to prevent future corrosion.

Chapter 4
Paint Stripping

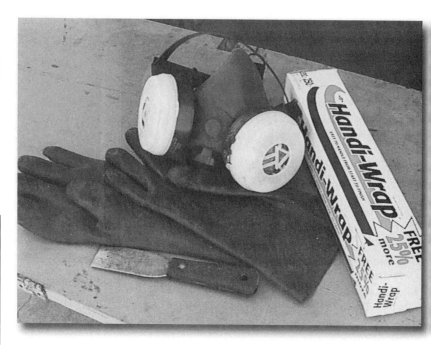

A charcoal filter mask and neoprene gloves are bottom-line musts for safety. A putty knife or Bondo spreader and sandwich wrap to prevent evaporation are important too.

Things You'll Need
- Paint gauge
- Aircraft paint stripper
- Goggles
- Propane torch
- Neoprene gloves
- Spray paint mask
- Cheap bristle paintbrushes 2"–6" wide
- 80-grit sandpaper
- Coarse steel wool
- Derusting solution
- Putty knives and Bondo spreaders
- Plastic kitchen wrap
- Lots of newspaper
- Cloth painter's tarp
- Spray painting rig
- DuPont's Kwik-Prep
- Epoxy primer

If you want to paint an old car properly, you may want to spend a few hours with a stripper first. Your spouse may not be happy about this and it can get quite messy, but there is just no better way to get paint off a vehicle. Of course, if a car only has its factory finish on it, you can just use that as a primer and scuff it to achieve a suitable substrate, but if the vehicle has been painted a couple of times, all the old paint will have to come off, and that's where the stripper enters the picture.

You see, paint that is too thick will crack. It can even alligator and flake off because a new coat of paint is only as good as whatever is under it, and may not be compatible in any case. Besides, each new layer only bonds to the one immediately below it, so you run the real risk of having adherence problems in any case. Any stress or vibration of thick paint will crack the finish, let moisture in and cause rust. The only answer is to take all of the old paint off and start over.

Paint Stripping Methods

There are a number of ways paint can be removed from a car. You can take your parts in and have a pro bead blast them using plastic beads and a media blaster, or you can take your car's parts to a Redi-Strip or similar professional paint stripping business in your area. You can even make your own

stripping vat using crystal drain cleaner and hot water. Or you can get a couple of gallons of Automotive and Aircraft Paint Stripper from you local automotive paint store or by mail order from the Eastwood Company, and strip the car yourself.

Any way you go about it, stripping all of the old paint off of your classic will be a lot of work. Fortunately, it involves minimal skill, and can easily be done by even a hobbyist, without much drama. But, safety precautions are paramount. Aircraft stripper is toxic. Stripper will blister your skin, and the methyl chloride in it will damage your lungs if you breathe it. Bead blasting makes a big mess too, and if you don't cover yourself carefully and wear a respirator, grit will get in your lungs, eyes and clothing and do serious injury. There is just no clean way to do this job.

Don't be tempted to try conventional sandblasting on your vehicle's sheet metal. Even when blasted with sand by a skilled operator, your parts are likely to come back warped, thinned, peened and roughened. And in the hands of a novice, sandblasting can be disastrous. The process can cut right through sheet metal and can easily damage panels beyond repair. If you take your parts to a media blaster, make sure they use something other than sand, such as fine glass, plastic beads or walnut shells.

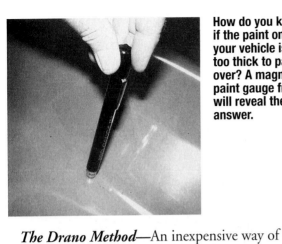

How do you know if the paint on your vehicle is too thick to paint over? A magnetic paint gauge from will reveal the answer.

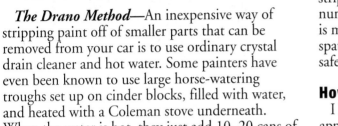

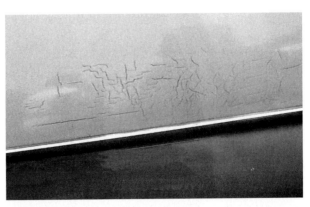

This alligatoring and checking is the result of a paint being incompatible with the layer underneath.

The Drano Method—An inexpensive way of stripping paint off of smaller parts that can be removed from your car is to use ordinary crystal drain cleaner and hot water. Some painters have even been known to use large horse-watering troughs set up on cinder blocks, filled with water, and heated with a Coleman stove underneath. When the water is hot, they just add 10–20 cans of caustic soda (lye)—also sold by the brand name Drano—and then start throwing in parts.

The process works in ten to thirty minutes on most parts and strips components easily down to bare metal. Afterward, all you have to do is rinse, dry, de-rust and paint. A tank like this will work for several days before it needs punching up with more lye or changing completely. And it can be poured down household drains.

This caustic brew will work with cold water too, but is quicker and more effective when heated to 100–120 degrees Fahrenheit. Be sure to wear goggles, neoprene gloves, and thick, protective clothing while working. If you splash any on you, rinse immediately with clean water. And never work with such a diabolical mixture around kids or pets. Those skull and crossbones warnings on the cans are there for a reason. This is deadly stuff.

Commercial Strippers—If you live near a Redi-Strip facility, you can simply take your car's body apart completely and have it dipped. Everything will come back clean as a whistle—paint and rust free—and with a protective coating. Then you just need to make repairs and start painting.

The only precaution you have to take is to make sure they rinsed all of the stripper out of every nook and cranny, because it will eat away at your new paint if you don't. Also, if your vehicle has any items on it made of nonferrous metals such as brass or copper, you will want to remove them before sending the part to the stripper, because such items will be destroyed by the stripping process.

Chemical Strippers—For most of us, probably the most practical and least risky method of

stripping a car for painting is to use any one of a number of liquid paint strippers. Although the stuff is messy and will burn your skin if it even lightly spatters on you, it is still the most popular and safest way to strip a vehicle of paint.

How to Strip Paintwork

I happen to live near a Redi-Strip, so my usual approach is to take bolt-on fenders, hoods, splash pans and the like, to them and have the parts stripped there to save time. Then I use aircraft paint stripper, which is a thick, nasty solution of methyl chloride that eats away the old paint to strip the rest of the car.

Remove All Brightwork and Trim—Remove any chrome, such as door handles, ornamental strips, emblems and badges. Aircraft stripper can turn chrome a light blue and cause it to lose its luster, and it will dissolve some plastic parts in no time. Wash the vehicle thoroughly to remove any grit before beginning to strip it.

Prep—Work outdoors on a calm day if possible. Good ventilation is a must. Put a large cloth tarp down to catch the drips. Also put on heavy clothing such as jeans and a sweatshirt, or coveralls. Finally, don a pair of safety goggles and a charcoal filtered painting mask. Don't be tempted to skip this step. Stripper is so nasty that it can damage your eyes, and your lungs if you breathe the fumes.

Sand It—The stripper will work a lot more effectively and quickly if you sand the car with eighty grit sandpaper to break the surface of the old paint before applying the stripper. The outer surface of paint is especially hard, and because it is smooth, it presents less surface to the stripper. If you sand the paint with coarse sandpaper, you increase the surface for the stripper to attack and you break through the hard outer layer of paint.

Take It All Off—Spray it on, in the case of the spray type strippers, or in the case of the Automotive and Aircraft stripper, use a cheap, throw-away paintbrush and spread a thick coat on an area about 3' square. Make sure your brush

1. Paint on the stripper with brush strokes only going one way. Don't brush back through your work. It hampers the stripping process.

2. Use cardboard and duct tape to protect areas you don't want stripped. Author painted firewall, then decided to strip and refinish the entire truck.

3. After the chemicals have had time to work (about 15 minutes), scrape off the bubbled paint with a putty knife or Bondo spreader. This is enamel. You know that because it wrinkles. Lacquer becomes soupy.

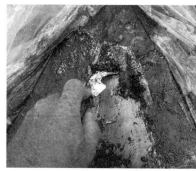

4. Keep a sloppy, wet coat of stripper on the area, and use sandwich wrap to retard evaporation on hot days. Paint should come off to bare metal. If not, apply more stripper and wait.

5. Plastic filler is ruined by stripper. Remove it, then make proper repairs.

6. A good professional de-rusting solution is a must before priming.

strokes go just one way, without going back through the stripper. That way you avoid hastening the evaporation of the most potent chemicals in the stripper. If you are working on a warm day, it is a good idea to cover the applied stripper lightly with plastic sandwich wrap to further retard the loss of active ingredients.

When the paint has bubbled and the stripper seems to have done all it is going to do—about 15 minutes usually—lift off the plastic and dispose of it in a trash can with a lid. Scrape the paint off carefully using putty knives or plastic Bondo spreaders if you want to avoid gauging the metal underneath. If the paint does not come off to bare metal, don't keep scraping. Instead shoot or brush on more stripper and let it set another few minutes before scraping again.

Wipe your putty knife or scraper onto old newspaper and place the dirty paper in the trash can. If your car is an older one from the '40s or '50s, it will most likely have been painted many times, in which case you may well not get down to metal the first time you apply the stripper. Work one part at a time until you have all of the paint off, then go on to another. Use strong twine to clean down into grooves such as drip moldings.

You can tell what kind of paint was on your classic by the way it comes off. Enamel wrinkles and crinkles, but lacquer turns to soup. Either one comes off pretty easily, but lacquer makes a bigger mess. It is always interesting to see the different colors and types of paint a car was painted with at different times.

If you don't finish your stripping job all in one day, rinse and dry the areas you have worked, and wrap them in cheap plastic tarps to protect them against moisture. This is important, because even in dry weather, bare metal rusts in a hurry. If you are going to have to store items for any length of time, you will want to shoot on a coat of epoxy primer.

Also, any areas you come across in the stripping process that contain plastic filler will need to have the filler taken out and replaced because stripper destroys it. You can grind it out with a grinder and a heavy-duty, twisted wire wheel, but a quicker, less messy way is to heat it with a propane torch to soften it, then pop it off with a putty knife. Just don't get the metal too hot and cause it to warp and oil can.

Clean It Up—When your parts are completely stripped, rinse them thoroughly with detergent, clean water and coarse steel wool making sure to get absolutely all of the stripper out of every groove and fold. If you get impatient here, any little bit of stripper left behind will eat away at your new paint and ruin your work. Dry your parts thoroughly.

Prepare It—Make any repairs required. It has been my experience that you can't really appreciate the extent of the repairs that will be required on an old car until you strip it. Paint can hide a multitude of sins, including rust and badly done previous repairs. When you have everything the way you want it, go over your parts with 80-grit dry sandpaper to develop a tooth to the metal. Finally, use a good metal prepping solution like Kwik-Prep to etch the metal and convert any unseen rust.

Paint It—It is important that you protect all the parts you have stripped as soon as possible. If you don't, you will soon have to contend with surface rust that will eventually bubble and eat through your new paint work if not completely removed. One good product for this purpose is DuPont's Variprime. Another excellent one is Epoxy Primer.

These are not thick surfacing primers. Instead, they are two-stage primers that bond with metal like you wouldn't believe. In the case of Veriprime, it actually etches into the metal. But make sure that any product you use will be compatible with the rest of the paint system you intend to use. These primers protect your parts from rust until they are ready for finishing, and they create a good surface for your final coats. Follow the instructions on the bottles or cans, and let your parts cure indoors for four or five days before going any further in order to release any gasses that might be present in the primer.

Keep in mind that, when you have finished stripping all the old paint off your classic, the worst is over. The next steps are final bodywork and metal finishing, then comes painting to produce that show-winning finish. When it's all done, you'll be glad you spent a little time with a stripper.

7. It is best to do bodywork before priming, but parts need to be protected until you can get around to doing the work, so protective priming may be required if you need to work over a period of time.

8. A good epoxy primer will protect your parts from moisture until they are ready to paint.

9. Pantera, stripped naked, will soon have a flawless gleaming red paint job thanks to Tom Horvath.

10. Work a small (three feet square) area at a time and make sure you strip to bare metal before moving on.

11. Even lead corrodes and crumbles with time. The rear of this vintage sprint car will have to be redone.

12. If a car only has its original coat of paint on it, you can scuff the original paint and shoot over it, but if the car has been painted more than once, chances are the old paint is too thick.

Chapter 5
Degreasing, Derusting & Prepping

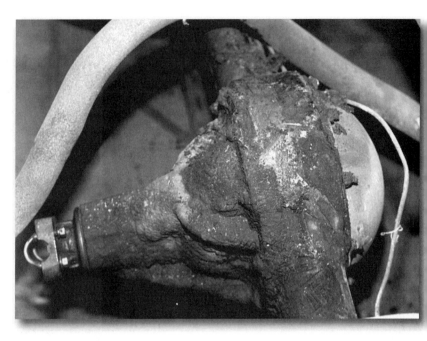

There is only one way to get this caked-on combination of dirt and gear oil off and that is with a scraper.

Things You'll Need
- Putty knives
- Small squirt can
- Lacquer thinner
- Grime Blaster from Eastwood
- Drill and assortment of wire wheels
- Strong twine
- Tri-sodium Phosphate (TSP)
- Scrub brushes
- Neoprene gloves
- Jackstands
- Safety glasses and paper particle masks

Body panels generally stay pretty clean on old cars, but fender wells, mud sills, floor pans and chassis components can get absolutely grimy in fairly short order if a car is driven every day. Dirt by itself is easy to get off, but when it combines with oil and grease, that's another story. Steam cleaning and media blasting doesn't phase the stuff I call grunge. (There is a very good cleaner with the brand name Gunk that works well to remove grunge by the way.)

Solvent will dissolve caked on filth eventually, but if you have to remove a lot of it, the residue will make a big, flammable, toxic mess. Then there is undercoating. You can't do bodywork on a fender until all the existing undercoating and dirt is removed from the underside, because you won't have a good surface to work on if you leave it in place. And media blasting won't touch undercoating either.

Media blasting works wonders on paint and rust, but all it does to grime and undercoating is heat it up and make it run and stink. When a professional blaster is all through, your components come out looking like dishes that you put in the dishwasher without removing the caked on food first. Most of the metal will be clean, but the residue from grime will still be there, defiant as ever.

Removing Grime

Grime and undercoating are not all bad though, because they do protect metal from corrosion, which, in the case of undercoating, is its primary job. It's just that both substances are soft and flexible, so they absorb the impact of even the most aggressive media blasting materials. Getting grunge off is easier than removing undercoating though, so we'll talk about that first.

Steam It—If you need to do the underside of an entire car, you can take it to a do-it-yourself car wash and use their steam gun to blast off grime. Don't do the engine compartment though, because you can easily damage electronic devices if you do. However, my favorite way to remove filth is to use a tool called the Grime Blaster. I hook it up to my water heater so I get hot water, then put the siphon hose into a strong mixture of TSP or even automatic dishwasher detergent. After that I hook up to my air compressor and go to work. The Grime Blaster will remove all but the most stubborn filth.

Scrape It—Next, put down some of those cheap throw-away plastic tarps available from building material and gardening stores to protect your shop floor. That way, when you are all done cleaning you can just roll them up and toss them in the trash. Set the car up on sturdy jack stands and make sure it is completely stable. Now gather a 1/4" electric drill, a few wire wheels, strong twine, a squirt can of solvent and some sharp putty knives. Put on full coveralls, eye protection, head covering and gloves and roll underneath the car.

Do as much as you can using the putty knives to pop off

Tom Horvath total restoration of a '56 Chev began with stripping down the frame. Note the meticulous authenticity of each component.

Eastwood sells a Grime Blaster that can be hooked up to your compressor that really cleans things up after you scrape off the caked-on grunge and undercoating.

When the chassis is finished, cover it with plastic tarps to keep overspray and dirt from settling on it.

After my truck chassis was stripped and cleaned we painted it with Eastwood's Chassis Black to get that OEM look.

caked-on filth. You can use the electric drill and wire wheel in places where the putty knife won't reach, but it is messy to use for general cleaning. Use the strong twine to get into grooves and crevices. If any areas are particularly hard and resistant, shoot them with solvent (paint thinner is good) and let it soak in before trying to scrape it off.

Remove the Undercoating

Water-based undercoating comes off with repeated wet coats of automotive paint stripper, but petroleum-based undercoating is nasty stuff to remove. It dissolves in gasoline or lacquer thinner, but both of these are highly volatile and toxic substances. Perhaps the safest and easiest way to remove it is to heat the panel from the back using a propane torch to soften the coating, then scrape it off using a putty knife.

Wear a respirator and work outdoors while removing undercoating. Keep a fire extinguisher handy too. Only heat the panel enough to soften the undercoating a little. You can easily oil-can and ruin panels if you get them too hot. Try just a little heat, then scrape immediately before the panel cools. You don't need to get every vestige of undercoating off this way. Just scrape off the thick deposits.

When you have scraped off the worst of the undercoating, come back with a rag soaked in lacquer thinner and wash off the rest of it. Wear neoprene gloves while working, don't smoke, and dispose of the oily residue according to local environmental laws. Keep the rag soaked with thinner and keep turning it as you wipe the area down. When you have the panel as clean as you can get it, wipe it down again with a good grease remover such as Eastwood's Pre.

After that, go over the panel with metal wash,

which further removes grease and contaminants and leaves behind a rust preventive coating that helps paint adhesion and prevent surface rust until you finish your bodywork and can paint the panel. Of course, you don't want to leave any panel unpainted for more than a few hours if possible, even after metal wash.

Popping Plastic Filler

The easiest way to get rid of large amounts of plastic filler (Sometimes called Bondo after one of the originators of plastic filler) is to heat it with a torch to soften it, then pop it out with a putty knife. The same admonitions apply here; for

Eastwood also sells all the authentic coatings to keep parts looking right too.

Eastwood's Vibratory tumbler is great for derusting and polishing hardware and small parts. Just add parts and water and turn the tumbler on. In a few hours the parts will look like new.

example don't overheat sheet metal, and keep a fire extinguisher handy. Small amounts of plastic filler can also be removed using a grinder and coarse grit abrasive wheels, but be careful not to damage panels if you go this way. If you strip your car with chemical strippers it will be necessary to remove all plastic filler too because the stuff will be softened and ruined by the stripper.

Blasting End Bolts & Fasteners

Anyone who works on old cars has spent time behind a wire wheel cleaning corrosion from bolts, nuts and other fasteners in order to be able to use them again. And if you do it long enough, chances are you'll have a bolt or two get loose and zing off. I know this because I've got a scar on my forehead as evidence of such activities.

However, these days there is a safer, easier way to do this tedious and somewhat hazardous task, and that is to put all your fasteners into a Vibratory Tumbler from the Eastwood Company and let it clean and polish them for you. The device looks like a motorized lettuce spinner, but don't let that fool you. It is very effective and easy to use. The kit comes with a supply of green rust-cutting media and white dry-shine media.

All you need to do is add the correct proportion of the rust-cutting media and fasteners and plug the thing in. How long it takes to clean up your bolts depends on how crusty they are, but within 24 hours at most they are ready to use. And if you want them shiny, you can put in some white dry-shine powder and tumble them again to really make them look new. The machine sounds a little like a vibrating bed, but the noise isn't unpleasant or intrusive.

These days, as soon as I get a car apart I degrease the fasteners and then pop them into the tumbler to clean them up while I do the rest of the work. Then when they are clean and shiny I give them a coat of paint or zap them with a little tin-zinc plating using another kit I got from Eastwood to give them that cad-plated look before putting them back into use.

Here's a tip: If you decide to paint or plate fasteners, be sure to tape off any threads so they won't get coated. Otherwise they may not fit when you try to install them. If you don't paint your fasteners, shoot them with a little WD-40 to keep them from rusting until you can install them again.

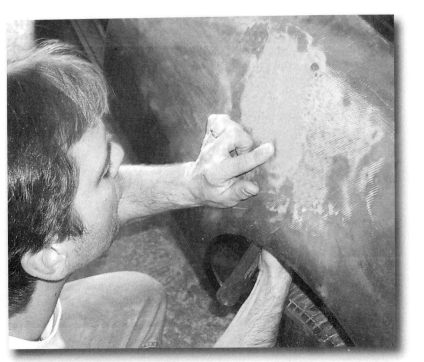

Tom Horvath locates a low spot with his finger so he can tap it in the right spot from behind using a pick hammer.

Chapter 6

Pro Hammer-and-Dolly Techniques

Metal finishing is an art. It's not like changing spark plugs. It takes lots of practice to do it right. That is the only way you can learn how hard to tap and where, in order to do the job. On the other hand, if you can drive a nail without bending it, you have the potential to do bumping and metal finishing. You will need the right tools and plenty of practice though. Body hammers and dollies (those little, handheld anvils used for backing and shaping sheet metal) are the main ones, and they are highly specialized. Their designs go back hundreds of years, in fact.

Tinkers, who were people who repaired pots and pans, first used them. Tinkers made their own tools, the designs of which evolved over time as the tasks changed. (Incidentally, the term tinker's dam refers to the clay dams that tinkers used to back holes they were patching so the molten solder wouldn't run out. Such a dam could only be used once—it was worthless after that—hence the phrase "not worth a tinker's dam.") Today, body and fender pros usually have a half a dozen or so hammers, several bumping dollies, plus body spoons and files, in addition to a lot of other more sophisticated tools to make their jobs easier as well.

You won't need that many items to get started, but as you continue to learn and earn, you will want to buy more tools. You'll need a minimum of three hammers, three dollies, a body spoon and a vixen file to start learning. The three hammers are: A combination hammer for general work, a long picking hammer for working in tight places, and a short picking hammer for taking out door dings. The dollies to start with are: a toe, a heel, and a general purpose. You can pick up these tools at your local automotive tool store, or order them from the Eastwood Company.

Choosing Dollies

Dollies that approximate but are not quite the same shape as panels are used to remove low spots. Use a flatter dolly for low-crown panels, and a rounder dolly, such as a general purpose dolly for high-crown surfaces. and the edge of a body spoon can be used to back hard-to-get-at grooves and fender lines.

Choosing Hammers

Picking hammers are good for getting into tight places, taking out small dents, and working raised lines in metal. A high crown, cross peen hammer is good for tight, concave, curves. With a general purpose dinging hammer, the larger head is for hammer-on work, and the smaller head is for hammer-off work. A short, beefy, heavy-duty bumping hammer can be used for roughing out dents, but a dolly usually works better for this job. Besides, a roughing hammer in the hands of an inexperienced person can do a lot of damage.

Technique

You will want to add tools as necessary, but first let's practice a little with what we have. Find a block of soft wood such as a pine to learn on. You can just pick up a short end

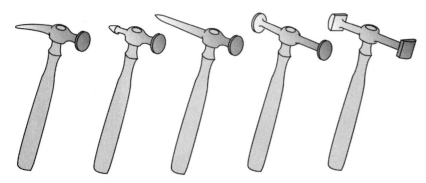

Here are some typical body hammers. From left to right: 1) Curved pick hammer, which is good for reaching into hard-to-get places, such as behind reinforcing braces. 2) Short utility pick hammer. Good for tight spaces. 3) General purpose pick hammer. This hammer is easy to use and has a good balance. Head has a low crown. 4) General purpose dinging hammer. The large face is used for hammer-on work. The small, high crown head is for hammer-off work. 5) High crown, cross-peen hammer. Primarily used for concave areas.

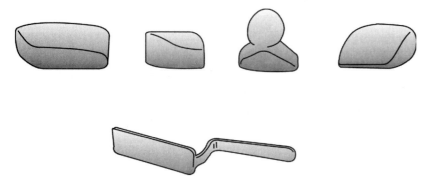

These are some common dollies: 1) Toe dolly. Flat face is used for working flat panels. Curved face is used for low crown surfaces. Also used for roughing out a dent. 2) Heel dolly. Good for reaching into tight corners. Can also be used for high crown panels. 3) General purpose dolly. Easy to hold, wide range of surfaces make this a useful addition to the bodyworker's kit. 4) Utility dolly. Good for high crown work. Edges are good for tight radii. 5) Body spoon. Used as dolly in tight places. Good for spanking dents out from behind. Also good to spread impact of a hammer blow on flat surfaces.

2"x 4" from a construction site for your introductory blows. While holding a general purpose dinging hammer loosely, about three quarters of the way up the handle from the head, practice tapping the block of wood using only wrist action and the opening and closing of your fingers to swing the hammer. Keep in mind that metal finishing is not bashing and pounding. Very little muscle is required to do it right.

If you are getting your arm into your swing, you're hitting too hard and you'll end up making dents in your classic instead of taking them out. Tap lightly, with the hammer head landing perfectly flat, so that you are not even denting the wood yet. Now increase your effort until the wood starts to dent. You should be making perfect circle

indentations that are no deeper along one edge than the other. Keep practicing until you can make clean circles routinely.

Now you're ready to practice on metal. Go to your local body shop or junkyard and talk them out of a damaged fender or panel that they will be discarding. Don't get a piece off of a modern Japanese import though, because they are thin and are made of high-carbon steel which is nearly impossible to work. The standard way to fix such cars is to cut out a panel and weld in a new one. And don't even think about just beating on your project car before you've practiced on junk parts until you are completely relaxed and confident about what you are doing.

Removing dents from body panels is not just slapping dented metal back into shape. It's more like unfolding a map. It has to be done in the proper sequence in order to avoid increasing the damage instead of fixing it. If you watch a pro, you'll see that he will take a fair amount of time determining the correct working sequence, and then get down and look at each dent from several angles before starting to work the panel.

That's because before repairing damaged automotive sheet metal you must first figure out how the dent happened. The reason this is true is because you need to work the dent in the reverse order of its occurrence. That means relieving the deepest primary impact first, and then taking out the interlocking cross ridges and valleys. Auto body panels are pre-stressed by stamping dies when they are made at the factory, and as a result, most of the warped metal in the dent will spring back to its original shape when the stresses caused by the interlocking creases and ridges are relieved by working the point of primary-impact first.

Broad Dents

The damage in our first photo of a Volvo fender at the top of the next page proceeded from back to front across a broad plane (A). As the car that hit our Volvo traveled forward, the fender was crushed inward and as a result, several deep channels were formed because the metal had no place to go. These creases are a lot like the wake a boat makes when plowing through the water. Amateurs would just start bashing the damaged metal around the edges of the dent from behind, hoping to batter the fender back to shape. All that would do would be to stretch the metal further and in unfortunate ways.

A pro would use a more analytical, and far more effective way to deal with a dent. First he would take the wheel off the car and carefully clean all dirt and undercoating from behind the dent. He would

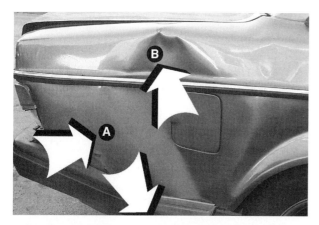

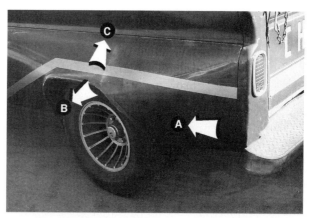

This old truck was obviously hit from behind (A) causing the fender to bulge, and also creating the ridges and valleys at (B) and (C). Pulling the bottom of the fender back will remove virtually all of the deformation.

Analyzing how a dent occurred is the key to removing it properly. Obviously this Volvo was hit from behind, causing the energy to travel forward (A) and causing the ridges and valleys (B and C) to form in front of it. If you pull the fender back at the rear, then work on the sharp crease at B, much of this dent will pop out without a lot of tapping.

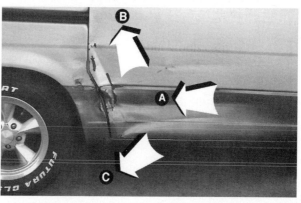

Something scraped along the bottom of this Aerostar from front to rear causing the ridges at (B) and (C). The bulkhead at the door interrupted the progress of the collision and caused more damage. This dent should be taken out in the reverse order of how it progressed. You would start where the primary damage is deepest at the bulkhead, and work out and toward the front of the van.

The dent on this BMW came from the front left. If you work on the crease where the major primary damage is, much of the dent will pop out easily because it is just stressed, not stretched.

Next, a pro would feel the dent with the palm of his hand to check for unevenness. After that he would use the hammer-on technique (The hammer-on technique means backing the metal with a dolly, then hammering directly onto it.) to take out high spots, then finish up with a picking hammer or bullseye pick to take out the small dimples.

Use your scrap fender or body panel to practice determining the direction of the dent and the area of primary damage. Start there and try to perfect your control. With the hammer-off technique you will see the dent start to rise. And with the hammer-on technique, when you are hitting squarely on the dolly, you will hear a healthy dinging sound.

Remember to wield your hammer only from the wrist and by opening and closing of your fingers. Your blows should start from a few inches away and should only rebound about a quarter to half an inch. The head of your hammer must land flat. Otherwise you'll only succeed in adding to your problems by making a series of little, half-moon or triangle-shaped dimples.

then remove the bumper. After that he would begin by slapping the rear of the fender back with a dolly of approximately the same curve as the fender in the opposite direction of the arrow A. In doing so, he would move the fender closer to its original shape, and he would probably pop out much of those interlocking ridges.

Then he would use what is known as the hammer-off technique to work out the rest of the dent. The hammer-off technique involves putting the dolly behind the valley and then tapping on the raised edges of the ridge with a body hammer. The dolly transfers the hammer's energy to the dent and pushes it out.

This is what happens when you just try to pound dents into submission without analyzing the dent first. All you succeed in doing is stretching and damaging the metal further.

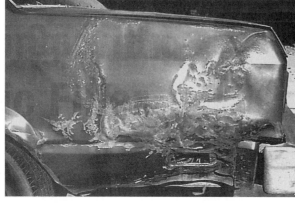

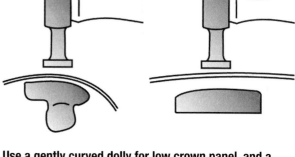

Use a gently curved dolly for low crown panel, and a rounder dolly for high crown panels.

The hammer-on technique is used to flatten smaller dents. The hammer-off technique is used to push out larger dents by transferring energy.

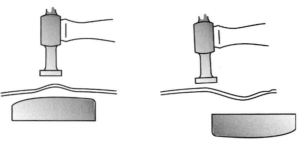

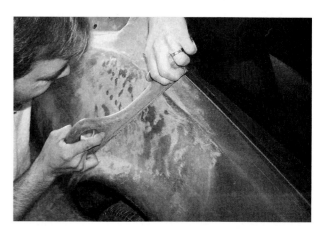

A vixen file is used to smooth irregularities and find low and high spots after the initial dent work is done.

Door Dings

Picking hammers concentrate a great deal of force on a small area of metal. They are used to take out low spots and door dings. It takes a lot of practice to get the hang of tapping directly onto the high point of a door ding. Try holding the tip of your finger over the dent to help you locate it. When you are pretty good at hitting the ding at its center, try backing it with a dolly. If you don't practice this technique adequately you will only succeed in making dings that stick out next to the little dents in the panel. Of course, a set of bullseye picks can make taking out door dings a lot easier.

Low Crown Panels

Use a body spoon to work large, flat, or low crown panels that are hard to get at from the back. A spoon can be used to reach in behind reinforcing braces to spank out low spots, and can be used directly under a hammer to help spread the force of your blows to take down high spots in low crown surfaces.

Use a fine vixen file to find high and low spots and keep working them until the panel is as level as possible. Don't overwork the panel though, because you will stretch and work-harden the metal if you do. Top pros can completely metal finish most dents to the point where no body filler is required, but most of us aren't that good. But if you have your panel to the point where filler not thicker than 1/8" is all that is required to even it out, you'll be fine.

A long flat lever can make a handy repair tool for use through holes in tight places, but be careful not to distort the panel.

Chapter 7

Limited Access & Paintless Dent Removal

There are a number of ways to do most things—some good and some not so good. In the area of limited-access dent removal, the new techniques and technology are vastly superior to the traditional approach. Tools make all the difference.

Often as not in dent removal, sheet metal doublers block your access, reinforcing braces stop you, and sometimes car bodies are made in such a way that you can't get behind their sheet metal at all. Several techniques have been used in the past to deal with lack of access, all of them bad. The first and worst was to clean up the area of the dent and then pack plastic into it until level. This was a terrible idea because plastic filler should never be more than 1/8" thick at its thickest. If it is any thicker it will shrink, crack and fall out. No self-respecting pro would ever consider this approach, but some botch shops do it to save money. They figure that by the time the car develops problems it will be a long way down the road.

Another bad approach is to drill a bunch of little holes in the dent and then pull it out using a slide hammer and hooks. If you aren't extremely careful doing this, what you end up with is a dent with a lot of little raised volcanoes in it. The next step is to smear on a thick coat of—you guessed it— plastic filler, being sure to push it into the aforementioned little holes so it will stay in place. A pro would weld up the little holes but an amateur won't usually bother.

Even though the plastic applied to fix the dent is thinner because of the raised areas, the filler oozes out of the little holes on the other side and forms little knobs. These are rarely sealed, so they wick up moisture, then the plastic falls out and the panel rusts. And, even if you weld up the holes, you still wind up with a weak, mutilated panel under the filler.

Of course, you can always cut out the damaged panel and patch in a new one, but that is a lot of work just to take out a small dent. And depending how you cut the area out, you can create major distortion that will have to be corrected. Then when you go to weld the patch in, the heat oxidizes the panel and makes it prone to rust. Heat also warps panels, and dealing with that can be a big challenge too. But thankfully, there are some new tools and techniques that make dent removal in tight places a lot easier than it used to be.

Tools

If you can afford one, an Arc Pull 'N Spot is a good way to go. It uses a spot welder and a lever to pull once the welder has bonded to the metal. The device is also capable of shrinking stretched metal. They're a bit pricey at $1,000 but if you can afford one, it will make dent pulling a breeze. Another advantage to the Pull 'N Spot is that it runs off of a car battery so it is totally portable.

Slightly less efficient but much less costly is a Stud Welder Dent Pulling System. It works by spot welding small two-inch-long pins to a dent, then pulling the low spot up with a slide hammer or dent-pulling tool that grips the pin. It is amazing how easily dents can be removed with these devices.

You can use Eastwood's Stud Puller dent remover to weld a stud in a low spot.

An Arc Pull 'N Spot is the ultimate for taking out small dents and is fully portable because it draws its power from a car battery.

Tom demonstrates the Arc Pull 'N Spot by welding on a stud, then pulling up on the lever.

Eastwood's Deluxe Stud Pulling System comes with everything you need.

Then special pliers or a slide hammer can be used to tug the dent out exactly where you want it.

Eastwood also offers a deluxe stud welder that is more versatile. It can be used with a shrinking tip to draw in stretched metal. You can also get a pliers-type dent puller that looks somewhat like a pop rivet gun and works well too.

Another device that does a great job in tight places is a slim pry bar that can be slipped under fenders and into door hinge openings to gently pry dents out. These have to be used carefully and you must make sure that you have a rigid item to pry against. Prying one piece of thin sheet metal by bracing against another may only make things worse and could ruin a panel.

A couple of old tools that can also be valuable for removing difficult dents are a plumber's helper and a bullseye pick. A plumber's helper—that rubber plunger with a wooden handle you use to unstop a drain—is great for popping out shallow dents when the metal isn't badly stretched. They come in various sizes, and I recommend you pick up one of each. Then all you have to do is wet the surface with a little soapy water and gently push the plunger on to the dent. A swift tug is all it takes to pull many dents level.

A bullseye pick is very accurate, though it does require a little space behind it. It is a great tool for door dings and small dents because you can hit them exactly where you want to strike them. Anyone who has used a picking hammer knows that even the best panel beater can't hit a ding square every time, and if you miss, you have a raised ding next to a depressed one. These devices come in three sizes. Those are the tools for creative dent pulling; following are some techniques that will help you do a good job.

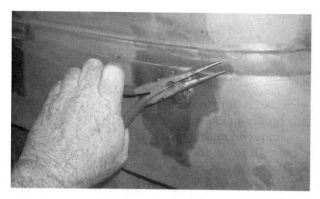

After the low spot is pulled flat the stud is clipped off and the bump ground down.

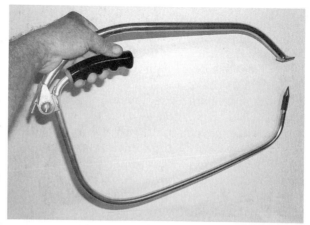

Bullseye picks are great for removing small door dings because they hit the spot every time.

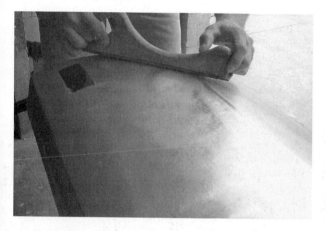

Often with this technique you can actually metal finish with no filler needed.

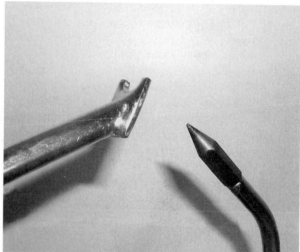

The U-shaped tip of the bullseye pick is placed on top of the dent, then the lever is used to strike it from behind.

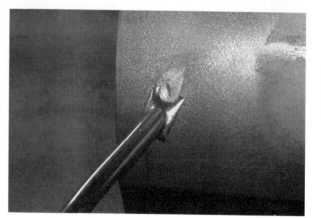

Often the bullseye pick can be inserted into hinge openings and other confined areas to take out small dents.

Techniques

Unless the dent is shallow and you can get it out with a dead-blow hammer or a plumber's helper, you will have to remove all the paint and any rust or plastic filler that may have been applied to a previous repair. If you are just doing a spot repair, you can grind the paint off with a sander and an 80-grit disk. If you are going to have to repaint the entire panel or the entire car, you may want to strip it with chemical stripper. Media blasting is another possibility but must be done with the right medium and shot by a pro to avoid peening, hardening and warping the panel.

Once the panel is stripped of paint, plastic and visible rust, use a sturdy twisted wire wheel to further clean the panel. A wire wheel gets down into the pores of the metal and does a much more thorough job of derusting than a sander, and it removes less metal. Only when you have the dent and the surrounding area spotlessly clean of rust should you proceed to remove the dent. Plastic filler as well as spot welders need clean metal to work right.

On the back side of the dent, if possible, scrape away any dirt or rust as well because dirt will cause any blows from behind to be uneven, and rust will just continue to spread and eat up your classic's sheet metal. Later, when the dent is removed, shoot a little undercoating or cavity wax up inside the panel to seal the metal.

Start your dent removal by analyzing the dent carefully. Look for the final point of impact. On very shallow dents it may not be evident, but with deeper ones you can generally see where the colliding object first hit and where it wound up. Start working the dent at the point of final impact. The reason for that is, often the surrounding metal is just flexed or distorted, and much of it will pop back into place with a little tapping or pulling with a stud welder.

Close your eyes and run the palm of your hand over the dent slowly. I know this sounds a little kinky, but the skin on your palm is quite sensitive, so you can feel even minor imperfections. Keep working the dent with whichever tool works best until you have it pretty well pulled out. Finally, go over the area with a fine vixen file so you can see the high and low spots. Work the low spots a bit more to try to bring them up, but don't overdo it.

A thin skim coat of plastic filler will last as long as the car does, and will remove any imperfections. If you overwork an area, it will become stretched and thinned. Of course, the initial collision may have stretched the panel too, and in either case it will need to be shrunk back to shape for it to look right.

The final finishing instructions in this chapter are different than you will find elsewhere in the book. They call for using epoxy primer, though it is just one good way to go. Another good way to finish a dent is to use polyester primer for final surface preparation, then go to final coats and color. That's because polyester primer is not as resistant to moisture, and should only be used if you are going straight to the finishing coats.

Once you have worked the dent as well as you can get it without overworking the metal, use a thin coat of plastic filler (no more than 1/8" thick before sanding) to level any slight inconsistencies. Top panel-beating pros can metal finish to the point where no filler is needed, but if you are new at the game don't try it. You could easily end up with a bulging, overworked mess.

As soon as the filler has hardened sufficiently—normally about half an hour—use a sanding board or air or electric sander and 80-grit paper to remove the excess and shape the repair. Sand from the edges in toward the middle, and sand parallel to the crease so the plastic will make an even transition from the surrounding metal to the patch. If you need a second application of the filler, make sure you mix it exactly as you did the first batch.

Filler that is two different hardnesses as a result of sloppy mixing will sand unevenly, leaving a bigger mess than the original dent. When you have the area shaped properly with 80-grit, finish shaping it with 120-grit paper. If you find any small pits or pinholes fill them with a little automotive glazing putty, but don't use it for anything bigger than pinholes.

Give the job a few hours to cure then give it a couple of coats of epoxy primer-sealer. After the epoxy primer has cured a few days (follow the instructions that come with the primer) shoot on a high-build primer surfacer. The final step is to block sand the area to remove any slight unevenness. The time you spend block sanding and final finishing your car or panel will make the difference as to whether your efforts will look spectacular, or will show unevenness under certain light.

Any way you work it, these alternative tools and techniques will give you professional repairs that will last for years, and they will not weaken or damage your chariot's irreplaceable tin. When you can't get behind your panels, it pays to get behind new technology to make your repairs.

Paintless Dent Removal

Most of us have heard of using plumber's helpers to suck out shallow dents, and we know that rubber mallets can be used from behind a small dent to take it out without ruining the finish, but here's a new system that can save you a lot of time and money, and is great for limited access dent removal too.

The kit is a bit pricey but it is the greatest thing since Bondo. It can pull out door dings and small dents with no damage to the paint at all, and it is easy to use. It works by sticking a small pull-rod to the surface of the dent with hot glue just enough to pop out the dent, but not firmly enough to damage the paint. Here's how to use it:

Work outdoors in a cool or shaded area. If the dented panel gets too hot it will affect the adhesion of the system's glue. Also, to prevent damage to your car's finish from cracking, the ambient temperature needs to be between 50 and 75 degrees Fahrenheit. Start by cleaning the dent thoroughly with silicone remover or alcohol. Ordinary rubbing alcohol will do the job. Use a clean cloth and let the area dry before proceeding.

Now take out the pull-rod and sand its tip lightly with 80-grit sandpaper so as to help the hot glue bond better. New pull-rods still have a residue of release agent on them from when they were manufactured, and you need to get that off so the hot glue will bond sufficiently to it. Otherwise the rod and glue will separate prematurely and not exert sufficient pull.

If you have a portable fluorescent light source such as a trouble light, position it at a low angle to the dent for maximum contrast so you can easily see what you are doing. Plug in the glue gun, put a glue stick in it and let it heat up. Next apply a small gob of the glue to the plastic tip of the glue rod. Immediately press the rod on to the exact center of the dent. If you wait too long, the glue will cure and the system won't work. Don't press hard because you will spread the glue over too large an area if you do.

Let the glue dry for one to two minutes. If you want to hasten the drying time, a blast of compressed air will help. It is very important to stay within the one to two minute curing time though because if you pop the puller too soon the dent won't come out, and if you pop it too late, you could lose some paint in the process.

Put the jig with the suction cups attached over the pull-rod and install the big wing nut. Turn it down until you feel tension, then turn the wing nut quickly until the pull-rod pops off the panel. It is important to shock the metal and not allow it to resist the pull. The suction cups on the jig can be moved in toward the dent for more pull, or out on the jig to give a little more cushion.

Remove the pull-rod or glue by using the glue remover. Be sure to let the glue remover completely penetrate under the glue to ensure easy removal of the pull-rod.

1. Here is the complete kit including glue gun, glue sticks, cleaner, jig, suction cups and pull-rods.

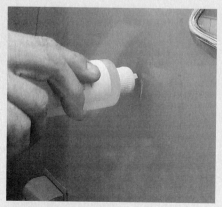

2. Use a little cleaner to get any silicone, oil or grease off of the dented area.

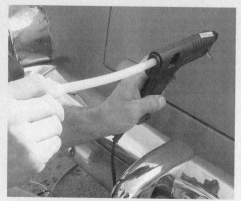

3. Slip a glue stick in the back of the gun, plug it in and turn it on.

continued on next page

continued from previous page

Once the pull-rod is separated, the glue will just peel away easily. These steps can be repeated as many as three or four times. Don't be overly concerned about creating a high spot because that can be leveled out using the included nylon punch by tapping the area gently.

The only caveat with this tool is that when using it on repainted surfaces you must be careful because the paint may not be as well-bonded to the car as was the original. If you have a little touch-up paint, you can try the dent puller in an inconspicuous area just to make sure you won't have a problem.

However, we used the dent puller on my old '57 Chev parts chaser, which has been resprayed and has since oxidized and aged, and it worked fine. The only place where we pulled paint off was where a little hand touch-up work had been done. The PDQ Paintless Dent Puller is available from www.pdqtools.com.

4. Shoot a gob of glue onto the tip of the pull-rod.

5. Stick the pull-rod to the center of the dent.

6. Attach the jig and suction cups, and then tighten the wing nut in the center.

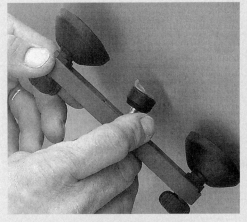

7. As soon as you feel tension, turn the wing nut quickly to pop out the dent.

8. If you need a stronger pull, slip the suction cups in toward the center of the tool.

A coarse file can be used to shrink slightly bulging areas simply by spanking the area so as to use the file's serrations to gather the stretched steel.

Chapter 8

Metal Shrinking

Automotive sheet metal becomes stretched as a result of collision, as well as when it is overworked. The result is bulges, broad, shallow dents and oil canning, which is what the old timers call the effect you get when you push on an old-fashioned tin oil can lid. It pops in and out because there is no place for the excess metal to go. It is even possible for a panel or fender to become so stretched and overworked that it is not savable, but most stretched panels can be fixed using these various techniques.

Spank It

Broad, subtle bulges can often be flattened using a coarse file. To take out minor bulges, spank the raised area with the coarse file rapidly so the teeth of the file gather the metal. This trick is good if you have only slightly overworked a panel and just need to flatten it a little bit. Make sure your blows are flat and hit the metal hard enough for the teeth to make an impression, but don't get carried away, because you will just make the situation worse if you do.

Dangerous Disks

I am only going to mention the wavy disk method of shrinking metal to warn against it. There are steel disks sold by a number of vendors that you chuck into a drill or die grinder and spin over the area that is to be shrunk. These devices generate a lot of heat and are supposed to cause the metal to tighten up.

They will shrink metal, but they are very dangerous because the wavy disk eventually fatigues and sends sharp steel shards zinging off in every direction. A number of unsuspecting body and fender pros have been seriously injured as a result of their use, so don't be tempted to try one.

Serrated Hammers & Dollies

If spanking the metal with a file doesn't do the trick, there are shrinking dollies and hammers. To shrink a bulge or dent where the metal is stretched, simply place a specially serrated shrinking dolly behind the metal and work it with a conventional hammer. Or you can use a shrinking hammer and work the metal over a smooth dolly. But never use a shrinking dolly together with a shrinking hammer. You'll damage the panel if you do.

The serrations gather the metal and shrink the panel, and if you use the serrated tool on the inside of a fender or panel, there is minimal finishing to do. If you must use the serrated face of a tool on the outside of a panel where it will be noticed, grind the area fairly flat, then use filler to fix any unevenness. (Don't get so aggressive with the grinder that you thin and heat the metal though.)

The Classic Torch Method

The most common and one of the most effective methods of shrinking stretched body panels is with an oxyacetylene or propane torch. Gather a bucket of water, cotton rags (no synthetic fabric because that could melt), a suitable hammer and dolly and a torch. Acetylene is best but you can use a

A shrinking hammer or a shrinking dolly can be used to gather stretched metal too. Never use them together though.

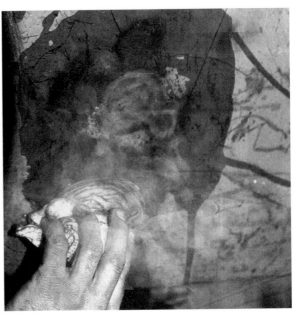

Quench the worked area with a cotton rag and water, or compressed air if you have it available.

Heat an area the size of a quarter on a stretched panel to cherry red using an oxyacetylene torch and a rose-bud tip.

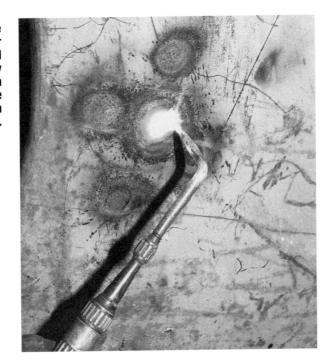

Quickly, while the metal is still hot, tap the bulging red area down using a hammer and dolly.

propane or MAP gas torch too. And if you have a good source of compressed air and a nozzle you can skip the water and rags.

Strip the area of the repair of paint and primer on both sides before starting to shrink the metal because the paint will burn, blister, make a mess, and cause toxic fumes. With an oxyacetylene torch, use a rosebud tip and adjust it to a soft flame that is lighter on oxygen and a little heavier on the acetylene. Too much oxygen will cause the metal in the panel to corrode and will make a very hot flame that could actually melt and deform the metal.

Heat a spot about the size of a quarter in the middle of the bulge to cherry red, then quickly place a dolly behind it and tap the hot spot with a body hammer. The heat softens and expands the hot spot into a distinct bulge. Then when you hit it with a body hammer the soft metal is forced out against the surrounding cold metal where it meets resistance. This hot metal has no place to go so it flattens and thickens.

After tapping the hot spot flat with a couple of blows, quench it with a rag soaked in water. This further shrinks and contracts the metal. But if you have a source of compressed air, you can give the hot spot a blast to cool it instead. This is actually a better approach because quenching hot steel with water gives corrosion a head start.

Shrinking just one spot may not be enough to fix a badly bulging panel though. In that case, work out from the center of the bulge and keep heating and tapping small areas. Just be careful not to overdo it. If—as in the case of a fender for example—you have another panel to compare with—keep checking to make sure both of panels end up with the same contour.

You will have to do a bit of metal finishing around the heated areas too, in order to make sure they are consistent with the surrounding surface. I have seen panels that—after having been painted— look as if they had a case of smallpox at some point. That's because there are all those recessed dimples that never quite got metal finished properly.

As soon as you have a panel shrunk to where you want it, carefully sand and wire brush off any rust, then shoot on a little urethane or epoxy primer to protect the area from corrosion. Panels that have been heated are much more prone to rust if not protected immediately.

Stud Gun Shrinking

Another easy way to shrink panels is using a stud gun equipped with a shrinking tip to draw in stretched metal. As mentioned in the last chapter, Eastwood sells a deluxe stud gun and accessory shrinking tip. Once the metal is clean, simply mount the shrinking tip in the welder, plug it in, and press it against the dent while holding down the trigger. As soon as a circle the size of a dime glows red, let up and put the gun down.

You can then work the area just the way you would with the torch shrinking method. Finish by quenching the spot with a wet rag, or better yet a blast of compressed air. Keep working in a spiral fashion out from the middle of the dent. When you have the dent close to where you want it, finish the job with a little hammer work and filing with a fine vixen file.

A stud gun from the Eastwood Company is a great tool for shrinking stretched panels. It is also useful for pulling small dents, as described in Chapter 6.

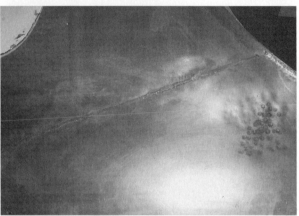

No matter which method you use, don't get carried away. You can end up making matters worse if you do.

Chapter 9
Basic Bodywork Welding

Here is what expert panel patching should look like: clean welds and straight panels with no warping.

Welding is not a dark art, but it is an art nevertheless. Learning how to handle an oxyacetylene torch took me a semester at the local junior college and I am still far from being an expert. In fact, to get a professional welding certification takes a couple of years of full-time schooling and a lot of practice using various types of equipment. But people who are into automotive painting and restoration don't really need to be expert welders. Luckily, with the advent of portable MIG and TIG welding rigs, you can learn to weld the mild sheet steel that car bodies are made of in just a few hours of practice.

MIG Welders

MIG (metallic inert gas) welding is a form of electric arc welding that uses a wire feed and a surrounding shield of inert gas such as argon, or an argon/CO_2 mix. There is also a special inner-shield wire that can be used with a MIG welder that creates its own gas shield as well. Inner-shield wire is not used for welding thin body panels though.

The thickness of the wire you need to use and the speed at which it is fed to the welding tip are determined by the thickness and type of metal you are welding. If you go with a Lincoln PAK 10 that operates on 110 household current, which is what I use, or one of Lincoln's more powerful PAK 15 welders that uses 220 volts you will find all the

My MIG PAK 10 was inexpensive and has served me well for many years.

information you need about amperage, metal thickness, wire feed rate etc. on the inside of the access panel on the side of the welder.

You will also need to rent a bottle of the inert gas from a

Play It Safe

1. Keep a good-quality fire extinguisher nearby. I cannot overemphasize this point. Anyone who has spent time welding has had the experience of working away on a project for a few minutes, then sitting back and lifting his hood to find that something nearby is blazing away cheerily. You can't see such a fire through your welding lens, so be careful. Also, never weld around gasoline, lacquer thinner, oil or any other flammable liquid, and never weld near a parts-washing tank.

2. Never weld wet. Make sure you, your rig and the floor are dry before turning on the machine. And never try to weld in rain or mist. This should be obvious but people try it occasionally. Even with 110 you could be taking your life in your hands. And even if the juice didn't kill you it would most certainly give you a jolt you wouldn't forget.

3. Always wear a cotton welding cap with a rear flap to protect your neck. These may be funny looking, but the first time a hot shard imbeds itself in your scalp you'll understand why they are a necessity. An ordinary cap turned around backwards to protect your neck isn't good enough. And if your cap is made of some sort of synthetic material a hot spark could actually cause it to melt, resulting in a serious burn.

4. Make sure you wear a full face mask with the proper level of eye protection at all times while welding. Those cool wraparounds you picked up at the last swap meet will do nothing to protect your vision. I've even seen welders look away and squint while welding. The damage this can do to your eyes is irreparable and you won't feel a thing.

5. Wear heavy leather gloves specially made for welding. Red-hot metal can sear your flesh in an instant. Also, let parts cool completely before trying to pick them up with your bare hands. Even thin sheet metal can take a few minutes to cool enough to be comfortable to handle.

6. Never attach the ground clamp to the bumper of your car and try to weld on a panel. Instead, attach it directly to the panel to be welded, as near as possible to where you will be working. If you attach it to your car's bumper your wheel bearings will become pitted, and any solid-state electronic components such as your on-board computer or stereo could be ruined by the high voltage passing through them.

7. Never attach your ground clamp to—or try to weld on—a gas tank, fuel canister or pressurized container of any kind. Gasoline permeates the metal inside fuel tanks, and fumes build up even when tanks are empty and have been stored for some time. There are professionals who specialize in repairing fuel tanks. You can usually find them at radiator shops. If you have a leaking tank, take it to one of them.

8. Make sure you close the valve to the argon or argon/CO_2 bottle when you are finished welding. These gasses are inert, but they can displace the oxygen in a room with potentially deadly consequences. Also, always make sure the gas bottle is chained to a welding cart or strapped to a wall so it can't fall over. Gas bottles have been known to take off like rockets and even go through brick walls if their valves are somehow snapped off when they are fully pressurized.

9. Always weld in a well-ventilated area. Some of the fumes given off by such things as galvanized iron while welding are toxic. Set up a large exhaust fan and work outdoors if possible.

10. Make sure you unplug your welding rig before changing polarity or changing rolls of wire, or making any kind of repair on the device.

local welding supply because these do not come with the welding rig. And while you are at the welding supply, pick up a pair of leather gauntlet gloves, a cotton welder's cap and a welder's leather jacket. This last item is not vital, but the first time a spark starts your shirt smoldering you will wish you had one. Finally, get yourself a good welding hood that covers your whole face.

With a gas welder you can get by with goggles, but if you try to weld with them using a MIG welder you will wind up with the mother of all sunburns. The MIG PAK 10 comes with a hand-held face shield that does the job, but a shield that fits over your head and can be flipped up and down is much more convenient. Also, if you can afford it, get a hood with a variable lens that darkens only when you are welding. These make life so much easier because you can't see anything through the correct number 12–14 lens when you are not actually welding. And yes, you can see through a

Leather gloves, a full face mask with proper lens and a cotton welding hat are a must.

The tank for the shielding gas is chained into the welding cart for safety reasons. Also note fire extinguisher on board.

number 8 lens in strong light, but that will not provide adequate eye protection for arc welding.

In our photos, Tom Horvath did some demonstration welding on steel coupons using a piece of aluminum as a backing. But a better way to go is to use a steel cookie tray filled with sand to back up your work. Fire bricks can be placed under the cookie tray for more protection. Just remember that steel generates temperatures in the range of 1,200—1,400 degrees Fahrenheit, so be careful where you weld.

Hot Tips

To begin welding, find some mild steel such as that in an old car hood or other panel. Don't use steel from a modern Japanese import though because the steel will be thin and very high in carbon, so you won't be able to weld it properly. Clean the panel thoroughly of paint, rust and grease. Never weld on anything but bright white

metal. Welds on dirty metal will not penetrate properly, will be hard to control, and will have a lot of inclusion of contaminants. A proper weld on clean metal is stronger than the surrounding metal, but a poor structural weld can be downright dangerous.

Never weld directly on a concrete floor either. The heat from the welder will cause the moisture in the concrete to turn to steam and explode, sending chunks of concrete everywhere. A steel or aluminum worktable with a vise is ideal for welding, but not really necessary for learning the basics.

Lap Welding

We'll do the easiest type of weld—called a lap weld—first. Cut your practice panel into six-inch coupons and dress the edges square. Lay two coupons out so they overlap by about 1/2". Hook up the ground cable so it grips both pieces and holds them in alignment. Check the thickness of your coupons, then adjust the wire feed rate and amperage to the specifications that came with your welder.

You are ready to try welding. Extend the wire at the tip to about 3/8" to 1/2". If it ends up longer, cut it off to that length using a pair of wire cutters. Touch the electrode to the work and depress the trigger. Just make a small spot weld, then reposition the ground strap out of your way. Tack a spot weld at the other end or your work to hold it in position.

Holding the nozzle at about a 45-degree angle, try to run a bead of weld between the two tack welds. It is very important to keep the gap from the nozzle to the work at the correct distance. You'll know when it is right because the sound of the welding will be sort of like that of frying eggs. If you hold the nozzle too far away from the work you will hear a hollow blowing or hissing sound. And, if you hold the nozzle too close, your wire electrode will stick in your work or to the nozzle.

Instead of watching the spark from the electrode while you work, watch the puddling behind the weld. The ridge of the puddle where the metal solidifies should be about 3/8" behind the electrode spark. I like to move the electrode from side to side a tiny amount as I weld in order to draw the weld in toward the center. Other welders move the electrode in tiny circles to do the same thing. However you do it, keep practicing until you get a nice clean-looking weld with good penetration.

Most first-time welders try to move too fast and wind up with ragged, weak welds due to a lack of penetration. Check to see if your welds are going all the way through by turning over the work. If a bit of molten metal has not oozed out and puddled on the back of your work, your weld has not penetrated enough. On the other hand, you may well find holes from melt-through too. These mean you lingered in one spot too long. Getting the penetration right is critical when welding on frames and structural members.

On very thin sheet metal, try drawing the electrode instead of pushing it. This can help give you better control. Keep working until you can produce a good strong lap weld before going on to other types of welds. Lap welds are easy to do, and very useful under certain circumstances, but the best types of welds for automotive sheet metal are butt welds. Lap welds mean that underneath the overlapping area is bare metal that will rust quickly if it gets any oxygen.

Butt Welds

To practice butt welds, place your coupons about 1/32" apart to allow for heat expansion and use your ground clamp or a C clamp to hold them in alignment. Tack the two coupons at each end as before, then once in the middle and let them cool for a minute. Next, practice running a bead between the tack welds. It will take a little more finesse to run a bead without melt-through when doing a butt weld. If the metal is very thin, only do about 1/4" at a time and let the metal cool between zaps so it won't warp.

When the panel has cooled, turn it over and check for penetration. Also, hold the test panel up to a strong light and look at it from the back to check for small pinholes. You would be a gifted beginner indeed if you didn't have penetration and pinhole problems on your first few efforts at butt welding. Keep trying until you can lay down a clean consistent weld that penetrates well.

Fillet Welds

Next, try a fillet weld, which is what you do when you want to joint two items at right angles to each other. Hold the nozzle at a 45-degree angle from vertical in order to weld on both pieces of metal at the same rate. Fillet welding is usually a little easier than butt welding though not as useful for bodywork. Move the nozzle back and forth as you work so as to knit the two coupons together. Again, stop frequently so as to let the parts cool and prevent warpage.

Plug Welding

Another kind of weld that can be very useful in automotive bodywork is the plug weld. Those little spot welds that auto manufacturers use to put cars

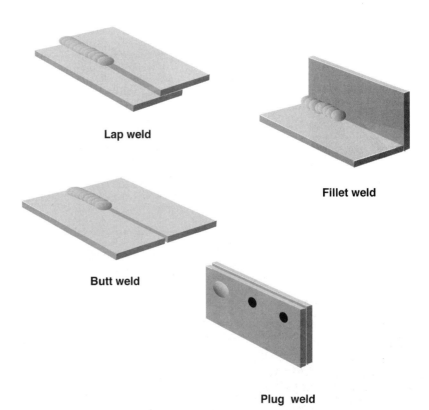

Here are the basic types of welds you will want to master. They each have their application.

Lap weld

Fillet weld

Butt weld

Plug weld

Tom's Technique

When working on the Aston Martins and Ferraris that came into Tom's Custom Auto Body, Tom never actually pulled a bead to weld thin panels. Instead he did a series of tack welds, making them ever closer together until he had a completely welded seam. This approach prevents warping and produces a very nice weld.

Tom Horvath tacks two coupons together with his bare hands. Don't try this at home.

I point to a typical running bead from a MIG welder. To the right is Tom's method using a series of tacks rather than a running bead. This makes a very nice weld.

To do a butt weld the pieces need to be about 1/32" apart to allow for heat expansion.

Note that electrode is about 3/8" long. If it gets longer, cut it off with wire cutters before welding.

Good penetration is a must for strong welds. Metal should be oozing out the back and there should be no pinholes.

Here is the sequence for welding sheet metal. Tack each end, then tack the middle. After that tack or run a bead a short distance from either end, alternating ends until you reach the middle.

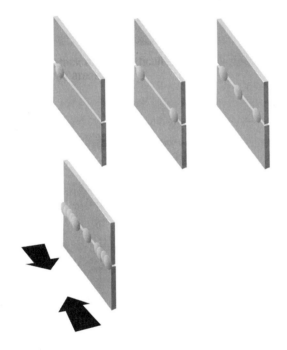

together sometimes have to be popped loose when panels need replacing. The Eastwood company sells a tool that will re-create spot welds exactly, but if you can't afford one, here's how to do a good imitation that will be as strong as the original:

All you have to do is drill 1/8" diameter holes at the same interval as the old spot welds, shoot a little metal etch primer on the facing flanges, then align the panels and fill in the holes using your MIG welder. It works like a dream.

Be sure though, as with all welds, that you clean the metal before and after the weld, then immediately coat it with a rust protectant such as epoxy primer, because welds rust much faster than adjoining metal due to the oxidation from the heat. The metal etch primer on the inner flanges will burn to some extent, but it will be better than bare metal.

One final word, though, is to keep practicing until you get it. HPBooks publishes *The Welder's Handbook* (see page 154 for ordering information). Also, even though it is seldom used for bodywork anymore, take a night class in gas welding if you can. There are occasions where you may need it, and you can also learn how to cut metal using a torch. And finally, if you are going to be working with a lot of aluminum, you may want to master TIG welding.

TIG is not as easy to master as MIG welding but it produces beautiful welds, and is great for welding aluminum. TIG welding also produces a softer weld that is easier to hammer and grind on. The only real drawback for many of us is the cost of the equipment, which runs about three times that of a MIG welding rig.

Chapter 10

Patching Panels

1. Fit your patch panel carefully over the old one and mark where the body needs to be cut using a magic marker.

Things You'll Need
- Replacement cab corners
- MIG welder
- Face shield with eye protection
- Leather gauntlet gloves
- Welder's cloth hat
- Heavy flannel shirt or leather welder's jacket
- Clamps, clecos or Vise-Grip pliers
- Die grinder or double-cut metal shears
- Waterproof primer

According to the guy who sold me my '58 Apache stepside it was a rust-free California pickup. I had my doubts. Then I stripped the old hauler of paint and plastic filler and discovered that it had the same rusted-out door hinges and rear cab corners that just about all working trucks develop after 30 or 40 years of service. In short, it was a mess. Lucky for me, some genius invented the MIG welder.

Order the Parts

We ordered new cab corners and hinge plates from Golden State Pickup Parts in Carson City, Nevada. They didn't cost much and Golden State shipped them UPS. They arrived in a day or two. We also picked up a couple of three foot square pieces of mild sheet steel at a local metal company to patch other holes, but if you don't have a place nearby where new tin is available, you can use metal from a junk yard donor. Be sure to get a panel that is the same gauge (thickness) and carbon content as the metal in your old truck though.

Gauge is easy to measure using calipers or a micrometer, but carbon content is a different problem. As an example of what I am talking about, if you are working on a classic American vehicle as we are, don't try to use panels cut from later Japanese imports, because they are made of thinner, higher-carbon steel. Instead, look for a hood or top from an old American car or truck. An easy way to verify carbon content is to hit the metal along an edge with a high-speed grinder. If it gives off sparks that are the same color and intensity as the metal in your truck gives off, you should be fine.

Critical Concerns

The two problems you can have that will really mess you up when welding in big sections like cab corners are heat buildup, and not measuring correctly to make sure the patch panel is properly aligned. To avoid misalignment, measure everything twice and take your time with fit. A little extra effort here will ensure a good job. We'll tell you how to deal with the heat from the weld a little later.

Cut It Out

Put the new patch panel up against the old area where it is intended to go and position it carefully so it is level and lined up properly. Also make sure it is not hanging down at the front or rear corners and that the lines are straight. Mark the upper edge with a magic marker as accurately as possible. It is okay for the patch panel to be a slightly different shape, provided it doesn't vary more than 1/16 inch out or in, because you can easily custom fit it using body tools.

To cut away the old panel you could use a high-speed angle grinder, a saber saw or even a hacksaw, but the best tool I've found for the job is a set of double-cut power shears. They give you a nice, clean line and don't warp and deform the panel. Never use a torch, and unless you really know what you are doing, don't try tin snips, because both can wreak havoc with your vehicle's tin.

Measure down 1/2 inch below the magic marker line you made along the edge of the old cab corner and draw a parallel line along which to make your rough cut. This will

2. Drill a hole using a hole saw so you can insert your double-cut shears into it, then cut to within 1/2 inch of your marked line.

3. Use the double-cut shears to trim and fit the opening to take your replacement panel.

4. Drill out the spot welds around the cab where the corner fastens to the inner sheet metal. Don't drill through the second layer if you can avoid it.

5. Also drill out spot welds in the door area and grind any roughness flat.

6. Use a wire wheel mounted on an angle grinder to clean rust and old paint from the area to be welded.

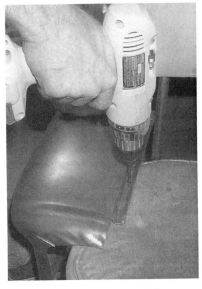

7. Drill the holes for plug welds before mounting the patch panel. Plug welds closely simulate spot welds and are just as strong.

give you room to do some fine tailoring. It is very easy to cut away too much and have a large gap, especially in curved areas.

Use an electric drill and a hole saw to make a hole in the old panel for the blade of your double-cut shears to drop into. Make your hole a few inches down from the line at which you want to cut so you can cut up to it in a sweeping arc rather that trying to make sharp turns and running the risk of damaging the blade of the shears. Now cut the old panel away along the lower line.

Spot Welds

Next you will need to remove any spot welds that hold the panel in place along such things as door seams. An electric drill with a 3/16-inch bit works well for this job. Just drill down into the weld enough to get the panel loose, but don't go through the bottom layer of metal. You may need to use a small hammer and chisel to pop the bad metal loose from the welds. Grind or file away any bumps that are left.

Make It to Fit

Cut a little away, then trial fit the panel, then cut and fit some more until you have a 1/32-inch gap all around, (1/16 inch maximum) and the panel's lines are level with the rest of the vehicle. Use a flat file to take way any rough edges and do your final shaping. If you make a mistake and make the gap a little wide in a couple of places, all is not lost. You can weld them up using a copper paddle held behind the gap.

Drill It

The simplest way to simulate original spot welds is to plug-weld a panel in place. To do that, first drill 1/8-inch holes in your new cab corner about the same distance apart as the old spot welds. Later, when you have your panel fitted and held in place, just fill the holes with weld. Do a neat job and when you are finished, the job will look as good and be as strong as the original spot-welds.

Fit It

Make sure the two pieces you want to weld together are clean, bright and rust-free metal. Trying to weld rusty metal will give very poor results. I like to use a wire brush mounted on a high-speed angle grinder for this job. Just make sure you use a wheel that is designed for 30,000 rpm and not one of the cheapies made for a 2,000 rpm electric drill, because inexpensive wire wheels will send bristles flying everywhere when chucked into a high speed grinder.

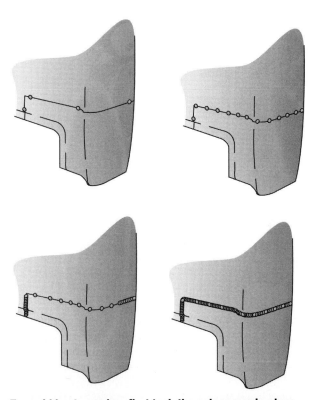

To avoid heat warping, first tack the cab corner in place at the ends, then tack it at 1 1/2-inch intervals working back and forth, letting the metal cool periodically. Finally, weld from one end then the other working back and forth between tack welds until you get to the middle. Use a die grinder to grind the weld flat, then you can use a little body filler if necessary to make the panel look as good as new.

8. Take a little extra time fitting in your patch panel. Make sure body lines are parallel and that the patch is contoured correctly. Use a body hammer and dolly to correct any problems.

10. A good weld looks like a stack of coins that has been pushed on its side. It takes a lot of practice, but with a MIG welder you can learn to do it pretty quickly.

9. Tack the panel in place at the ends and in the middle. Then tack every inch or so, letting the panel cool in between. Finally, run a bead between the tacks.

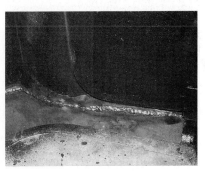

11. A good strong weld penetrates completely to the other side. Shine a light from behind to check for pinholes.

Once you have your patch fitted to the existing panel, Cleco them into place or clamp them with Vise-Grip pliers or "reach-arounds." Clecos are slotted clips that hold a panel in place and in line and help maintain the 1/32-inch gap needed for proper welding. Reach-arounds are specialized Vise-Grip pliers available from welding supply stores that will hold panels together too.

This is your last chance to easily adjust fit and alignment, so give the whole patch area a good look before lighting up your MIG welder. You sure don't want to find an alignment problem after you're through welding. And keep in mind that a minimum 1/32-inch gap is critical to allow the welded metal to expand from the heat without pushing together and warping. When you have everything just right, it is time to tack the panel in place.

Weld It In

Tack the patch in place near each end of the cut, then tack it in the middle. You are not in a hurry, so let the part cool between welds. Now tack back and forth, alternating from one end to the other about 1 1/2 inches apart. Finally, fill in between your tack welds, alternating between the ends of your seams to minimize heat buildup. Use Tom Horvath's method of building up spots (explained on page 40) rather than trying to pull a bead, so as to avoid warping thin sheet metal.

To minimize the amount of filing and grinding you will have to do, you can hammer weld if you can get in behind the panel. Weld about an inch, then place a body dolly behind the weld and hit it with a body hammer to flatten and force the molten metal out. You won't be able to make the line look perfect, but with minimal grinding you will produce a very nice seam that can be made invisible with a little filler.

You can also use a body hammer and dolly to

realign the heated panels as necessary while welding. Stop and inspect your work regularly. And look over your work from behind periodically to make sure your welding is penetrating properly. When you complete the weld, shine a light behind your work to check for pinholes if you are new at this.

Finish by putting in your plug welds, then let the area cool. MIG welding produces a very hard weld, so you need to work carefully while grinding it flat so you don't thin, heat up or warp the existing metal. Use a high-speed angle grinder to take down the welds, but rather than thin the metal

unnecessarily you can smooth the area over with a thin coat of plastic filler.

Remove any flashover rust or discoloration using your high-speed wire wheel again. Now use a good metal prep solution to clean and etch the new patch, then shoot on some moisture-proof epoxy primer to protect from corrosion until you are ready to paint it. Welds are especially prone to rust due to all the heat involved, so a certain amount of oxidation is inevitable.

Filling Small Holes

People drill holes in old trucks to graft on accessories and decorative items. Smaller holes can be easily filled using a MIG welder and copper paddle, and larger ones can be filled using the usual patching method. Here's how:

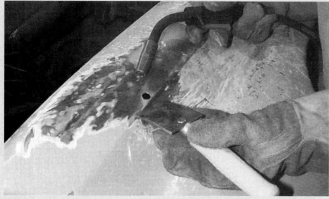

1. For small holes, just hold the copper paddle (available from the Eastwood Company) behind the hole and weld it up, working in circles until it is closed.

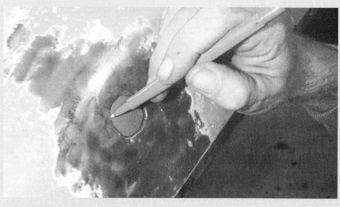

2. For bigger holes, such as this one used to mount huge signal lights on the fenders of my truck, first mark up a piece of sheet metal held behind the hole to make a patch.

3. Tack the patch in place while backing it with the copper paddle, then weld it in.

4. Grind it flat with a surface grinder, then give the patch a coat of epoxy primer to prevent rust.

Chapter 11

Door Skinning

This door suffers from badly done repairs and having its lower flange bent in to try to get it to mate up with the sill.

Things You'll Need	
• Door skin stamping for your car	• Chisel
	• Electric drill
	• Cold galvanizing compound
• Hammers and dollies	• Mig welder
• Die grinder	• Cavity wax
• Cushioned work rack	• Primer and paint

Good friend and master craftsman Bruce Haye, owner of ATR Auto Restorations in New Zealand, was about to re-skin a door on an MGB GT when I dropped by to visit him. The fact that the car is British and the panel work was not done in the United States doesn't mean anything because the process is the same the world over. It's just that I had never seen the job quite so beautifully done before. Here's what it takes to do a door skin right.

Door skins are available for many older cars—especially popular classics such as Mustangs and later Chevs. In fact, if your older car's door is dented and bent, check parts sources on the Internet, club publications, old car parts purveyors and even dealers to see if you can find a decent door skin before resorting to such barbaric measures as slide hammers and hooks to get the dents out. These days, more and more inexpensive stampings are being made in Asia for just about any car available.

The problem with using a slide hammer to get at door dents you can't reach from the inside is that you have to drill holes in the door in order to use the hooks. Then you have to weld the holes shut to prevent moisture getting to the plastic filler you will need to use to further smooth the panel. The resulting welds make the panel more prone to rust, and besides, the filler will not hold up with the constant opening and closing of the door.

Turns out the original problem was with the sill, not the door. Bending the door was a mistake.

Reskinning doors looks easy—and it is—sort of. If you have never done bodywork though, I would leave the job to a pro, because little mistakes can mount up and make a mess of the whole thing. However, if you are pretty good with a hammer and dolly, you can do the job yourself and save some money.

Quick, even sweeps with a die grinder are the key to removing old door skins.

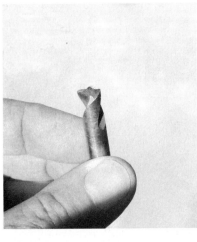

To drill out spot welds without drilling all the way through, a special bit is available.

When the lip starts to change color you are nearly through the flange.

Spot welds must be drilled out for outer skin of door to be removed.

Our '67 MG had been restored once before, and the job was done rather badly. When the previous panel beater got though welding in new rear fender arch patch panels and sills, the car's doors stuck out past the sills. He then tried to hammer in the edges of the doors to make them fit, and that was a big mistake, because doing so merely creased the doors along their lower flanges. Then the fellow compounded the problem by using plastic filler to reshape the bottoms of the doors. And the plastic filler started chipping off along edges every time anyone closed the door.

It didn't take Haye long to figure out that the problem wasn't with the doors but with the sills. And as luck would have it, not only were new door skins available for MGBs but sills and just about every other patch panel anyone could need were offered too. He ordered new sills, along with the new door skins, and they arrived the next day. Bruce checked the new sills against the old ones and figured out the problem. The inner sill where the scuff plate usually sits was made too shallow on the repro sills, so when they were welded in originally, the doors stuck out at the bottom.

It must have been the same problem the previous panel beater had, except he didn't notice it until he had welded in the sills and then it was too late. Bruce had to cut the new repop sills lengthwise and made new flanges so the sills could stick out where

they were supposed to be. He then welded the new flanges into place and welded the sills in carefully. Haye tells me that poorly fitting stampings are not uncommon, and the professional panel beater has to deal with them appropriately.

Only after fixing the sills could he attack the doors. And as it turned out, along with the creases, the original outer door skins were in pretty sad shape due to crudely done previous repairs that had been slathered with filler and painted. Bruce unbolted the driver's side door and put it on a cushioned work rack. The next task was to grind around the edges of the door to remove the old skin. If you are going to skin a door yourself, make relatively quick, even passes with a die grinder to avoid taking off too much in one sweep. Also, don't linger in one spot because you will create a notch if you do.

Keep making sweeping passes until you see the metal begin to change color. It will start to turn blue as the lip of the door skin gets thinner. At that point one last light pass should expose the

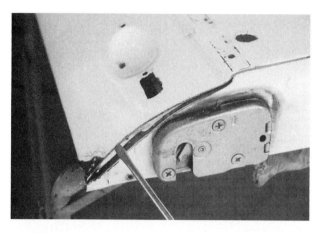

Lower lip of door skin just pops off with a screwdriver after the edge is ground away.

Mount the door with its skin off to see how it fits. That way you'll know where you need to make adjustments.

Door panel lifts away to reveal a rusty flange that needs to be cleaned and straightened.

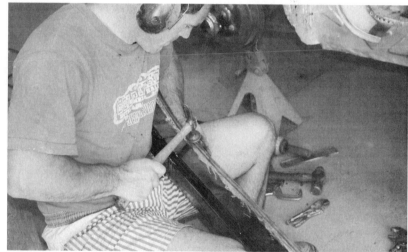

Master panel beater Bruce Haye extended the sill by making a new flange.

laminations of metal. Use a chisel or large screwdriver to pop the bent-over flange off. It will drop down pretty easily once you have cut through the edge of the door.

Next, you will need to break loose the spot welds that help hold the door skin in position. On our MGB, there were just four. Two of them were near the top of the front lip of the door, and two more were at the rear lip near the top. A special drill bit made specifically for drilling out spot welds will do a neat job of breaking the welds, or you can just use a regular drill bit if necessary.

Once the door skin is popped loose, use a hammer and dolly to take any ripples out of the door flange, and to make sure the flange is at the correct angle all around the door. When you have the flange as you want it, give all of the bare metal a coat of cold galvanizing compound to prevent further rust. Cold galvanizing can be welded over with minimum damage, unlike paint.

The next step is one you will hear again and again,

Place the new skin on the door and check to see if there are any deviations from original.

In order to do this job right, you must check your work frequently for fit.

Place the dolly along the lip and incline it toward the edge. Tap the flange back to about 45 degrees.

Sharp corners need to be rounded slightly and sealed so as not to cause problems.

and that is: Check the fit before going further. Bruce did this many times all through the door skinning process and it paid off because when he was finished, the door fit perfectly, and the gaps ran parallel all around. Mount the door back in place with no skin on it so you can see where its shape deviates from the opening it is supposed to fit into.

To crimp the door skin over the doorframe, use a hammer and dolly, and hold the dolly along the edge at about a 45-degree angle while you work. Just bend the lip of the door skin over evenly at about a 45-degree angle for you first pass, then check your work again to make sure everything is going to plan.

Never try to bend the lip over with the dolly back away from the flange edge because you will damage the door skin where it meets the lip edge if you do. On our MG door there was one sharp corner, and Haye flattened it with a couple of taps so it would not have a sharp edge.

Again with the dolly along the lip, tap the door skin flange flat against the doorframe. Work your way around the door evenly until the flange is over tightly and evenly. Now mount the door in place and check the fit again. Once the door has been carefully mounted, if the front gap is a little bit wide and the rear gap is tight, gently tap the door skin forward a little. And of course, if the door is tight up front after careful fitting, you can tap the skin rearward along its edge.

In the case of our MG door, the skin fit tightly at the center of the rear gap, but was slightly out at the top and bottom. Bruce gently tapped the end of the door in the middle and worked the lip down, then flattened the flange against the door again, thus shortening and flattening the curve where the door was tight. Then he carefully worked and flattened the flange in the area where the door gap was too wide, thus pushing it out toward the gap ever so slightly. This is the stuff that can take years of experience to get right.

Check the gap all around again, then take the door back off and place it on your work rack. Once you have the gaps exactly as you want them, drill the door skin near the original spot welds and weld the skin to the flanges underneath. Grind the welds flat. Mount the door again and make any final

Red arrows at top and bottom show wider gap. Arrow in the middle shows tight area.

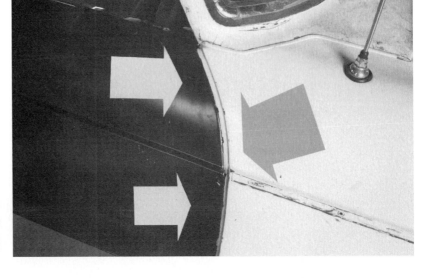

Wide gap at top may be because door is not hanging properly, or because of differences in shape between skin or frame.

adjustments. Look everything over carefully, then do any last hammering work to get the door exactly as you want it. No plastic filler should be necessary, nor is any advisable.

When you have the door as you want it, take it off again, clean and paint the inside, then use cavity wax to seal around the edges and wherever there were welds. Don't give rust a chance to get started again. Grease the latch and window regulator hardware and then mount side window in new whisker moldings. If all you were doing was skinning a bad door, at this point you could scuff, prime and paint the door to match the rest of the car. But Bruce is doing a complete restoration on this GT, so painting will have to wait until all of the other imperfections are dealt with.

Bruce Haye makes door gaps absolutely right by carefully working flanged edges.

Chapter 12
Adhesive Panel Repair

Jagged speaker holes were cut in the doors of my '58 Chevy pickup. These seemed like good places to try automotive adhesives.

Things You'll Need
- The Eastwood kit, which includes: flanging, dimpling and compression pliers; blind holders (20 will do for patches 6"x 8" or smaller); and automotive adhesive
- Tin snips, saber saw or double-cut power shears
- Pop rivet pliers
- Rivets (#4 steel countersunk)
- Caulking gun
- Files, body hammers, dollies
- Filler spreaders
- Lacquer thinner

It is the rare restoration candidate indeed that doesn't need a skin graft or two. The traditional way to do it is to MIG weld in new metal. And the best technique is to butt-weld in the panels because there is no way to seal overlapping flanges when welding in patch panels with lap joints, so rust forms rapidly.

Welding takes skill, and of course, a rather expensive welding rig. Thankfully these days there is a low-cost alternative, and that is to glue in patch panels. Don't laugh. I know it sounds shoddy but Chrysler and the European automakers have been gluing cars together for years.

The bonds made by the space-age glues these companies have developed are actually stronger than the surrounding steel. In fact the only way you can loosen adhesive-bonded panels is to heat them to 400 degrees Fahrenheit. The reason car makers are using these new glues is because the thin, high-carbon steel panels in modern cars rust and warp quite easily when heated and welded.

The only catch with adhesives is the panels need to be clamped together for at least 12 hours, and that isn't always easy to do unless you have the right tools. But as it happens, the Eastwood Company sells a complete kit that contains everything you need, including a tube of the adhesive to clamp panels anywhere on a car, so I sent for one of them. Even though I have a MIG welder, I wanted to give the new adhesives a try.

The kit works very well as it turns out. Gluing in a patch is a little more labor-intensive than welding one in, but it requires no special skills, and best of all for those of us

working on a budget (and who isn't?) no welding rig. I decided to fix a couple of my '58 Chevy truck's doors that had jagged holes cut into them for speakers. They came out beautifully.

Strip the area to be patched down to bare metal. A sander and 80-grit sandpaper disk is good for this, or you could use chemical strippers, then hand sand the area with 80 grit if you don't have a sander. Make sure you strip at least two inches beyond any rusted or damaged metal because you will need to overlap your patch by 5/8 inch, and you will be using the glue to seal the edge as well as the lapped seam.

In the case of my truck doors, those jagged, amorphous holes were too funny a shape to try to match, so I drew out a rectangular patch with a pencil and straight edge, then drilled 1/4-inch holes at its corners. With that done I started my double-cut sheet metal shears at the speaker hole and made the cuts. You could use a saber saw or sheet metal nibblers, or if you are experienced enough, you could even use aviation tin snips to make your incisions.

Clean up the resulting edges with a file or angle grinder. If there are any warps or bumps resulting from the cutting, use a hammer and dolly to get the metal to the shape you want it. In our case that was easy because the metal of the doors was flat along a horizontal plane, though there were some corrugations in it.

Cut your new patch out of steel that is the same gauge as your vehicle's panels and ideally the same carbon content. (Carbon content is absolutely critical when welding in patches, but not as important when using adhesives.) If you use appreciably harder, stiffer steel than the surrounding metal in an area that will flex, any paint and filler you put over the patch could crack along its seams.

To determine the correct size and shape for the patch, place a piece of stiff bond paper over the opening and rub along the edge with a pencil. Next, add two inches all around to your template. Finally, place it on your new metal and cut it out carefully. The patch will be about 3/8" too large all around allowing you to do a little custom fitting and still maintain the correct overlap.

In the case of my Chevy truck, our new patch had to be contoured to the existing panel. I used an English wheel for this and just rolled in the corrugating along the lower part of the panel. You could make similar contours without a wheel by shaping a piece of soft pine to the contour you need, then gently hammering the metal into the shape using the plank as a buck, but a wheel makes metal shaping so much easier and more professional looking. After I rolled in the curves I used a 1/4-inch rod and a block of wood in a vise to press in the ridges in the patch.

Fitting and lining up the patch is critical to the final appearance of the fix. Make sure grooves and crowns line up, and that you have a minimum of 5/8" overlap along all the edges. Cut and file your patch to the shape you want, but don't get too fussy, because you will need to make further adjustments after the edges of the opening are flanged.

In order for your finished patch to be flush with its surroundings you will need to create a recessed flange around the opening that is just the right depth for the new panel. Flanging tools are available that can do a very nice job, or you can just use the reworked Vise-Grips that come in the Eastwood kit. Setting the tool so it clamps crisply and makes the proper impression is easy.

Make sure the flange is parallel to the edge of the hole and lined up on the inner edge of its die, then just squeeze down. Keep working your way along the hole until you have the edges done. We chose to round the corners of our patch to make the flanging look neater and more consistent, but you could also work the corners of the flanged hole with hammer and dolly to get them just right.

After you have the hole flanged all around, check your patch again to make sure it fits and is flush with the surrounding surface. A little snipping and filing should take care of any problems. Finally, smooth and deburr the edges of the patch with a file. If any flattening needs to be done, place the patch on a wooden surface and gently tap it with a body hammer.

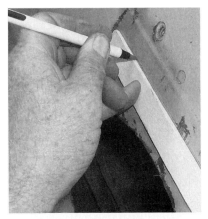

1. Draw out a rectangular shape with a pencil and straightedge.

2. Drill 1/4-inch holes at the corners to help avoid over-cutting.

3. Use hammer and dolly to make cuts completely flat.

4. An angle grinder makes short work of cleaning up edges, but a file will do the same job.

5. An English wheel made forming the contours of our panels easy.

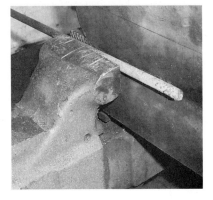

6. We used a steel rod and a block of wood to bend straight lines in our patch.

continued on next page

continued from previous page

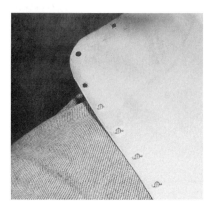

7. Use a file to deburr the edges of your patch and the holes for the rivets.

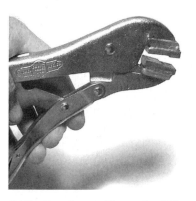

9. The flanging tool is a pair of Vise-Grips with special jaws for recessing the edges of patches.

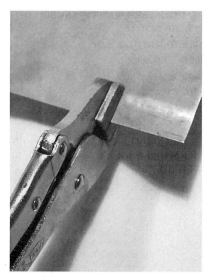

11. Line the flanging tool up along its back edge and clamp down to recess a flange into the surrounding metal.

8. Carefully align your patch, then drill a 1/8" hole and install a holder. Do the same on the other end of the patch to locate and stabilize it.

10. Assemble the glue cartridge and mixing tip, then insert them into a caulking gun.

12. Dimple all of the holes you drilled in the panel to be patched and the patch panel itself.

Place your patch on your bench and measure in 3/16" all around and make a line with a pencil or magic marker. Now mark off spots for holes along this line. Space your marks 3/4" apart on patches of 5"x 3" or smaller, and 1" apart on larger patches.

Put your patch into place in the flanged area and drill a 1/8-inch hole through both surfaces. Install a blind holder using the compression pliers. Now drill a hole in an alternate corner of the patch and install another holder. With these in place, drill the rest of the holes through the patch and the flange.

Lift the patch out of the way and deburr all the holes on the flange and patch to ensure a flat fit. Next, use the dimpling pliers to countersink the holes in both the panel to be patched and the patch itself. Finally, use a hammer and dolly again to make sure both surfaces are flat. The dimpling process can make minor distortions that will make the patch stick out if you aren't careful at this stage.

Put the patch panel back on and install the blind holders so you can do a final test fit. If there is any unevenness, you may have to use the dimpling tool again to make the countersinking a little deeper, or you may have to employ hammer and dolly to get things perfectly flat. Patience and care at this point will save you a lot of filing and shaping later and will ensure a secure bond for the two pieces of metal.

Remove the blind holders and clean the area thoroughly with a rag and lacquer thinner. The adhesive will establish a tight, durable bond only on paint-free, rust-free, and grease- and oil-free metal. You would need to do the same if you were welding in a patch, so this step is almost universal when making any kind of body repair.

At this point it would also be wise to mask off the bonding surfaces on the back of the patch and shoot on a little moisture-proof primer or rust inhibitive paint to keep the patch from corroding. Just make sure you leave the area to be bonded completely clean and bright.

Next, take the cap off of the two-part adhesive tube and install the plastic mixing tip. Place the adhesive tube in the caulking gun and place the blue adapter-spacer behind the glue cartridge. Squeeze a little of the adhesive onto a scrap of cardboard to make sure it is mixing properly. The glue should appear uniform in color with no streaks in it.

Carefully apply the adhesive to the original panel in a 1/4"-wide bead along the centerline of the rivet holes. At this point you have about 30 minutes to position the patch before the glue sets up. Install the patch panel using blind holders in every hole. When that task is completed, if the panel you are patching is flat, use a straight edge to make sure the panel is flush. But don't worry about squeezed out adhesive just yet.

Instead, remove the blind holders one at a time and install countersunk steel rivets. Work from the centers out to the corners, alternating back and forth across the patch to minimize stresses and buckling. Throw the blind holders in a can of lacquer thinner as you remove them, and wash them clean of adhesive quickly to avoid having them glue themselves together.

Finally, use plastic filler spreaders to spread the excess adhesive. Once the adhesive has set for four hours it can be sanded, filed and worked as you would conventional plastic filler. You can also use regular plastic filler for fine finishing and to fill any slight irregularities. Just be sure to let the adhesive cure for 24 hours before doing so. Finally, shoot on a little protective epoxy primer until you can get around to painting the part.

The patch will not rust and will hold up for as long as the surrounding sheet metal. And with the usual finishing work it will look as good as any other patch. Also, you run no risk of distorting and oil-canning your classic's priceless tin from excessive heat. Give gluing a try. You'll be surprised at how well it works.

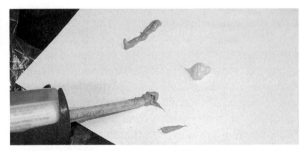

13. Make sure the adhesive is mixing thoroughly as it comes out of the tip of the gun.

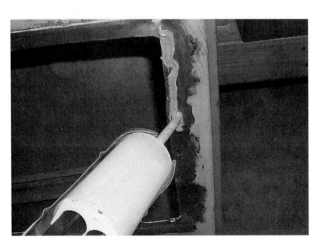

14. Squeeze a 1/4" bead of adhesive all around the opening, then position the patch carefully.

15. Use compression pliers and blind holders to clamp patch in place.

17. Replace each blind holder in turn with a countersunk rivet to clamp patch in place permanently.

19. Let your patch cure for 24 hours, then finish in the usual way with plastic filler.

16. Clamp your patch temporarily with blind holders in each hole.

18. Spread adhesive with Bondo spreaders to smooth it. Let it set up for four hours, then sand it to shape.

20. Finish with a coat of moisture-proof epoxy primer to prevent rust until you are ready to paint.

Chapter 13
Metal Fabricating Tools

The MetalAce English Wheel is perfect for the home hobbyist because it takes up very little room and can even be chucked into a vise.

I used to write a column for *Auto Restorer* magazine called Tool Talk in which I reviewed new or particularly interesting tools and advised people where they could obtain them. I wound up with a lot of tools that way, and only reviewed the ones I thought were truly useful. And out of all these columns I have chosen a couple of tools that I especially like and feel every good panel beater should have.

English Wheels

I have spent too many hours tapping on pieces of mild sheet steel over a shot bag to rough them into shape, only to wind up with rather lumpy replacement panels that sort of fit with a little tweaking and fiddling. I'm sure those Italian masters that made the custom bodies for Ferraris by hand back in the '50s would laugh themselves weak at my humble efforts, but—with a bit of plastic filler—my panels did the job.

Thankfully that's all behind me now because I was able to find an inexpensive English wheel that can roll smooth, beautiful curves into sheet metal in a few minutes. And though everyone calls these devices English Wheels, the one I obtained is actually made in the U.S. by MetalAce and is much less expensive than imported models, though its quality is outstanding.

The large, flat anvil wheel, as well as the roller, have heavy-duty needle bearings in them for durability, and both are precisely machined and polished. MetalAce makes huge, heavy, professional English metal shaping wheels as well as smaller, hobbyist's wheels that will handle up to 20 gauge sheet steel and can be clamped into a vise such as the one I have, plus a number of models in between. And all of them are available by mail order directly from the manufacturer.

It takes about five minutes to install the anvil wheel and the pressure adjustment crank before you are ready to go to work. Grease the bearings on the wheels before you begin using them. Rollers will wear out eventually, but MetalAce offers replacements for them. Otherwise, the tool is maintenance free. Wear leather gloves while metal shaping because you can easily get your fingers pinched, especially while attempting to roll a curve for the first time.

The hardest part about using an English wheel is getting the tracking right. At first the sheet metal seems to have a mind of its own, but after half an hour of playing around with a piece of scrap sheet metal you do start to get the feel for it. It seemed much like learning to back up a trailer for the first time. The metal wants to go the opposite direction from what you expect. Of course, it takes a fair amount of practice to become proficient at shaping panels, but it's well worth the effort. Just keep in mind that the object is to use the wheel to gently stretch the metal into the shape you are looking for.

Wear heavy leather gloves while working because the wheel can pinch your fingers.

The author's initial efforts were rewarding. The wheel put a nice even curve into our scrap piece.

Wheel pressure is governed by a crank at the bottom.

Only use light pressure while learning, and don't use a great deal of pressure even when you've mastered the machine. The wheel's rollers are capable of over a thousand of pounds of pressure, and will actually cause bulging furrows in the sheet steel and ruin the roller and upper wheel if you get carried away. The trick is to work the metal back and forth slowly under fairly light pressure, overlapping as you go, sort of like mowing a lawn. If your rolls are too far apart and zigzag all around, you will distort the metal in ways you'd rather not.

When you have practiced enough to control the tracking well and can control what you are doing, you are ready to try making a patch panel. Make templates from cardboard or make a buck to establish the curve you want, then go to it. Clean and strip the sheet metal of paint, rust and grit. Just as with hammers and dollies, if there is grit on the metal you are working, it will make dents in the rollers, and those will then emboss themselves into your work.

If you do decide to get a wheel for your shop, you might also want to pick up a copy of the *Metal Fabricator's Handbook* by Ron Fournier. It tells you everything you need to know to fabricate almost any panel you might need for your restoration. It is available from HPBooks (see page 154), or you can order it on the Internet from Amazon at amazon.com, or online performance parts retailers jegs.com, and summitracing.com.

Cool Cutting Tools

One of the most critical steps in patching body panels on old cars is cutting out rusted sheet metal. It's one of those jobs that definitely takes the right tools. Common methods for cutting classic tin such

Rollers are available in different radii for different curves, and can be changed easily.

Double-cut power shears are really the way to go when cutting sheet metal.

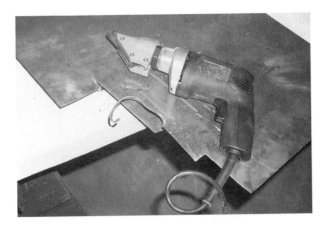

A good alternative to double-cut shears are sheet metal nibblers, but you need a healthy air supply to drive them.

Double-cut shears give you a clean straight line that makes welding in new panels much easier.

Wiss aviation tin snips are also a must, and come in rights, lefts and straights.

as using die grinders or torches leave jagged, nasty edges, and both approaches heat and warp the panels. The metal remaining is then brittle and prone to rust, and must be worked back into shape. Thankfully there are much better ways to do the job these days.

My favorite method is to use electric, double-cut metal shears. These things are marvelous because they can cut around gentle curves and they make a clean, straight line without bending panels or causing jagged edges. They are also easy to control. Before I discovered double-cut shears I used to use a die grinder, which made a fairly straight but nasty cut and left uneven, scorched edges.

I've also used saber saws, which are hard to

control, but generate less heat. The biggest problem with saber saws is the vibration they set up. It can bend and tweak the tin and cause an irregular line. Another problem with saber saws is the depth the blade extends while it is jumping around. If there is another panel close behind the one you are cutting, the blade will hammer against it and cause problems. A saber saw can also damage any wiring behind the panel you are cutting.

With double-cut shears you can cut through one panel at a time, and you really only have to do a minimum of dressing of the edges of your cut with a file to get new panels to fit properly for butt welding. Quite a lot of filing and fitting is usually necessary when you use die grinders, saber saws or cutting torches.

Welding panels in place is so much easier if you have a nice clean, straight, edge with a consistent 1/32" gap. There is minimal distortion due to expansion and a minimum of burn-through due to a wide gap—both common problems for beginning welders. A ragged, inconsistent edge makes for a ragged, inconsistent weld and a lot more metal finishing or plastic filler to make the job look right.

I got my double-cut shears from the Kett Company. They sell retail for around $200 depending on the thickness of metal you want to cut, but they are well worth the cost. If you are going to need to fix extensive rust—and sooner or later you will, if you are restoring cars—double-cut shears will pay for themselves quickly by saving you a lot of grief and drama with warpage.

If double-cut metal shears are beyond your budget, another very good alternative is a sheet metal nibbler. I picked up a compressed–air powered nibbler at a swap meet, but Campbell Hausfeld makes better ones that do almost as nice a job as the double-cut shears. The only catch is that you need a good compressed-air supply to power them. If your compressor and tank aren't large enough, you'll be stopping every couple of inches to let the air build up before you can continue. The device will just run out of air and bang against the metal. I also discovered that the nibblers run into problems wherever there is a double thickness flange. In that situation, A keyhole or saber saw can help. Or, if the flanges are tack welded, you can separate them by drilling out the weld, pulling the reinforcing panel out of the way and cutting each piece separately.

Sheet metal nibblers work by punching little holes in rapid succession. Nibblers are so fast in fact, that they cut a nice, straight line. They are an excellent alternative if you do have a healthy air supply, and they are long lived. About the only thing you need to do is change the little round tip periodically when it gets dull.

Other cutting tools that are indispensable when removing rust or fabricating new panels are aviation tin snips. You will want both right and left cutting snips so you can cut curves nicely. To cut with tin snips you snip, then push the snips in and up slightly before snipping again.

I use Wiss tin snips because they are high quality and hold their edges well. The red-handled snips are left hand cutters, and the green-handled ones are right cutting. There are also neutral snips (they have yellow grips) and these can be handy for straight cuts, though you can also do straight cuts with right- or left-hand snips. Wiss tin snips retail for about $18.00 and are available from most tool stores.

Sources

English Wheel
MetalAce English Wheels
75 Truman Road
Pella IA 50219
1-800-828-2043
1-641-628-8886
E-mail: sales@metalace.com
www.englishwheels.net

Cutting Tools
Williams Lowbuck Tools Inc.
4175 California Avenue
Norco CA 92860-1769
1-909-735-7848
www.lowbucktools.com

Chapter 14
Working with Plastic Filler

Grotesque misuse of plastic filler has given it an undeserved bad name. Lead won't handle this kind of abuse either.

Things You'll Need
- Professional quality plastic filler and catalyst (Evercoat Rage Gold is good.)
- Plastic spreaders
- Metal or plastic mixing tray (Never use cardboard. It will absorb critical chemicals and weaken the filler.)
- 80- and 150-grit open coat dry sandpaper
- Sanding boards
- Metal prep
- Polyester primer
- Lacquer thinner (one gallon) for washing
- Disposable facemask for dust and particulate matter

There are a few purists who are horrified at the thought of using plastic filler (often referred to as one of its early brand names, Bondo) to finish a collector vehicle for painting. They claim that lead is all that was ever used before the mid-1950s and that it is more flexible and more durable than plastic filler. They are right on all counts. The only question is, how durable and flexible does body filler need to be?

Plastic isn't strong enough to fill holes or patch along the edges of doors or hoods, or to be used in places that undergo a lot of flexing. Obviously, if you were going off road you would want to use filler sparingly, and never in areas that were subject to bending and torquing. But anywhere else—assuming you only use a thin coat and that it is protected—filler will work very well indeed. In fact, plastic fillers work so well that amateurs routinely abuse them, which is how they got a bad rap.

Some neophytes do nutty things like applying filler an inch thick over unprepared, painted or rusty dents, or filling large, ragged holes backed with nothing more than chicken wire or cardboard. And as if that weren't enough, some of these same people drive around for weeks and leave their cars out in the elements before giving the filler a protective coat of paint. Is it any wonder they have problems? Lead won't take that kind of abuse either.

Actually, plastic filler—properly applied—holds up well. So well in fact, that vehicles restored with it thirty years ago—though driven regularly—still look flawless. It really comes down to this: If you are doing a total cosmetic restoration on a vehicle during which the body must be stripped to bare metal, you will need to replace all the existing body filler—including that which was applied at the factory—with something, whether it be lead or Bondo.

Both lead and plastic deteriorate over time. So you need to decide whether you want to put up with the toxicity, expense, and learning curve involved in mastering lead just to get that last ounce of authenticity (even though you'll be the only one who knows) and that last word in durability for a car that will likely never be treated roughly again, or you are willing to settle for plastic filler, which is much easier and cheaper to apply.

It's a matter of choice. Lead takes a lot of practice, but you can do it if you try. On the other hand, if you decide—as have most pros—that plastic filler is suitable for your needs, read on. The technique to apply it is quite easy to master, involves no torches that can warp precious panels, and produces no toxic fumes or dust—though it is not a good idea to inhale the particles produced by sanding filler.

When dent is repaired as well as possible, sand off all rust and scale to bright metal.

Plastic filler works so well when applied properly that it is even being used to fill pits in this classic Rolls being readied for Pebble Beach.

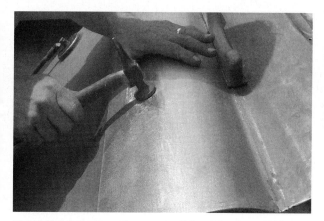

Wire wheel gets into pores of metal and removes last vestiges of corrosion.

Metal-finish dents as far as possible without stretching and thinning metal before applying plastic.

Metal Finish It

Neither plastic nor lead were ever intended to fill deep dents. The best body and fender pros only use either filler to smooth minor imperfections. If possible, they tap out, shrink and file dents to their original shapes. That is why you will want to polish your skills in metal finishing before plastering on plastic.

The less proficient you are as a panel beater, the more you will need to resort to filler of some kind. And if you are a hobbyist working at home, there is no question that you will need to use filler in places unless your project needs very little work, or you are exceptionally talented.

Clean It

When you have a dent worked out to the point where you only need a thin film of filler to make it right, you are ready to use filler for its intended purpose. One mistake beginners often make is to overwork dents in an effort to metal finish like the big boys. Don't do it. A little properly applied filler is not a problem, and is far better than ruining a panel by beating it until it is so thin, hard, and oil-canned that it can't be saved.

Before you open the can of filler, sand or strip any paint and rust off of the area to which you will be applying it. Use a good rust remover and follow it with a stiff wire wheel. This will etch the metal to give it "tooth." Only apply plastic to perfectly clean, bright, metal or a coat of polyester primer if the plastic filler allows. Wipe the surface down with a good metal prep solution such as DuPont's Pre-Kleeno to remove any silicone that has permeated it as a result of waxes applied to the car over the years, but never use any kind of acid prep because filler won't stick to acid treated metal.

Mix It Up

Before you mix up a batch of filler, massage the tube of catalyst with your fingers to mix it and warm it. Also, stir the filler thoroughly, because it separates out in the can in storage. Now mix up only as much filler as you can easily spread in about ten minutes. That is specified as a line of catalyst across a 4" diameter gob of filler with Evercoat, but follow the mixing instructions on the product you purchased. Filler sets up to the consistency of Parmesan cheese in about twenty minutes so any

Massage the tube of catalyst with your thumbs to warm it before adding it to filler.

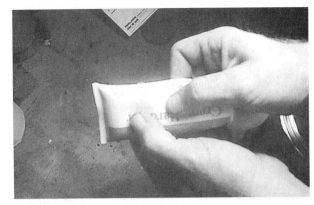

Mix filler in a figure-eight motion until color is consistent, and be very careful not to create bubbles.

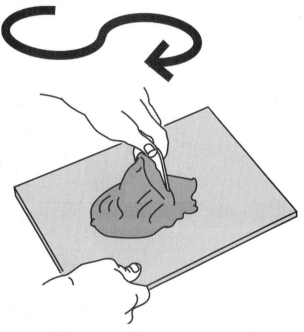

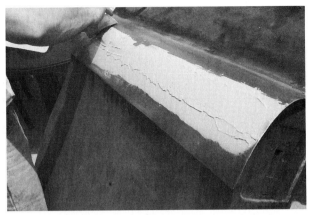

Tom Horvath quickly spreads plastic evenly along dent using a scrupulously clean spreader.

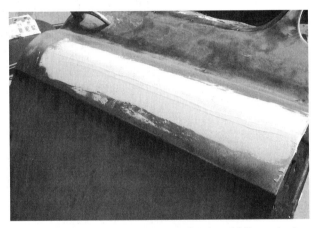

Filler is applied low at edges and high in middle so dent can be shaped and contoured to original shape.

Lay a strip of catalyst across a four-inch gob of filler and mix until color is consistent, with no streaks.

more filler than you can apply in about ten minutes will go to waste.

Mix the filler without pulling the spreader out of it and never chop it. Just work it around in a figure-eight fashion until there are no streaks left, and the filler is a consistent color. Don't mix beyond this point though, because the filler will start to set up immediately and become useless within a few minutes.

Don't try overdoing the catalyst or applying heat in an effort to get the material to set up more quickly. Both tricks have adverse effects on durability. Worse, don't try using less catalyst on a hot day either, because you will wind up with a sticky glob of goo on your car that will never harden. Mix the filler and catalyst thoroughly so your filler has a consistent color all the way through. Be careful not to trap air bubbles while you are mixing because they will cause problems later when you do your shaping work.

Apply It

Apply a thin, even, coat of filler no more than 1/4" thick which will be in excess of twice as thick as you want for a finished surface. The surplus is to allow for shaping and sanding. Keep your spreaders and mixing palette perfectly clean, because little bits of grit or hardened filler will ruin your work by making grooves in it as you attempt to spread the filler. Now, while your plastic work is hardening, is a good time to quickly wash your tools in lacquer thinner. If you neglect to do so, you will likely have to throw your tools away once the filler hardens.

Put on a particle mask for this next step. When the filler sets up to the consistency of parmesan cheese, shape it roughly with 80-grit sandpaper in a sanding board. Sand along the length of the dent in long strokes and work from the edges in toward the center to feather the filler and prevent low spots. Wherever possible, use a power sander or the longest sanding board that will fit. (Don't use a cheese grater file because it will tend to bruise the filler and cause it to shrink in unpredictable ways.)

Often as not, a second application of filler is required to finish the job properly. That is why it is important to follow the mixing instructions as precisely as possible. If you don't, your second batch of filler may be harder or softer than the underlying one, which will cause problems when you try finish sanding your work. Pros refer to the result as a fried egg. That is because poorly mixed filler often shows up as a sort of circle within a circle that looks somewhat like a fried egg on the side of your car after it has been painted.

Go to the 150-grit paper to do your final finishing. Glazing putty can be used to fill tiny pinholes, but don't use it for anything else because it is not strong enough. If you are unsure about whether you have your panel exactly the way you want it, shoot on a fine mist of paint in a contrasting color using an aerosol can, then sand it off using a sanding board. High spots will show up immediately, and low spots will continue to be speckled after several sanding strokes. Fix any imperfections before going to the next step.

When you have everything exactly right, let the filler cure for several hours in a dry, warm place before priming it, in order to allow gasses to escape. Finally, shoot on a little more waterproof polyester primer to seal the panel and prevent rust.

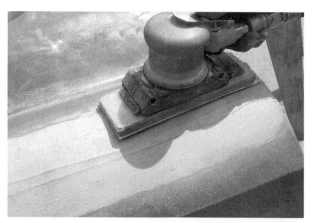

After plastic sets up, electric sander and 80-grit paper are used to shape the dent. Tom sands parallel to the major axis of the dent so it will come out even.

Final shaping is done with 150-grit paper and electric sander. Sanding boards will work just as well, but will take longer.

Chapter 15
Working with Lead

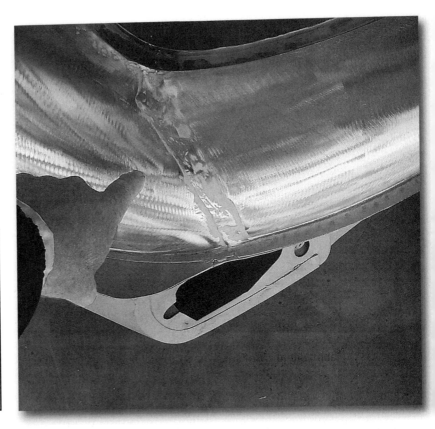

When Bruce received the Jag replacement panels from the U.K. he discovered that they all had downward bent flanges to fit under existing panels. This made fitting them together a challenge.

For years I thought that doing bodywork with lead was too old-fashioned and too difficult to bother with. But I've since learned I was wrong. In the past, I managed to get around it by welding in patch panels and using plastic filler. But recently I learned how to body solder, and found out why it is necessary in some cases. And I became a convert in the process. Lead work isn't as easy as spreading on plastic filler, and lead sticks can get expensive depending on how many are required, but there is no mystery to it, and no fancy tools are needed.

Of course there is no compelling reason to use lead in many cases. Plastic filler is great for filling low spots and smoothing unevenness in body panels, and it will hold up very well if applied properly. But lead is the best way to go on panels that will be flexing, or in situations where you need to create corners, and in places where body panels have to be joined in finicky compound curves, or where the filler will need to be thicker than 1/8". (Plastic filler should never be more than 1/8" thick after sanding.)

Here's a good example of a situation where lead is mandatory. My friend Bruce Haye is busy turning a Series 3, E-Type Jaguar coupe into a roadster for a customer. This conversion has been done with good results in the past, but it takes a craftsman like Bruce to do the job. It also takes several rear panels—some of which are available from the U.K. and some of which need to be hand shaped on a plenishing wheel. And as if that weren't enough, the panels from overseas are hardly precise fits, and they are not really designed to work with one another.

Replacement rear fenders, aprons and the like are made to be installed as individual patch panels for repair purposes but are not designed to be used to craft the whole rear of a car, so all the flanges along their edges are bent to go under adjacent, existing bodywork. That's fine if you are only installing one panel, but in the case of our Jag, it meant Bruce needed to mate two similarly bent flanges together.

He was able to do so successfully, but the result was rougher than just tucking one panel under the other. Lead was the only answer to finishing the job properly. Plastic filler would have had to go on too thick in order to get the contour right, and besides, it would most likely crack out as soon as the trunk lid was slammed a few times.

If you want to master the skill of applying body solder (leading) yourself, practice first on a scrap fender or door.

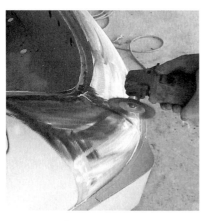

1. Before attempting lead work, a panel must be spotlessly clean. Here Bruce uses a wire wheel and a high-speed grinder to get into every little crevice.

2. Next, a sander and 80-grit sandpaper are used to remove any corrosion and to develop a tooth for the lead to adhere to.

3. When the panel is perfectly clean, tinning paste is brushed on. It contains tin and lead and is toxic, so be careful with it.

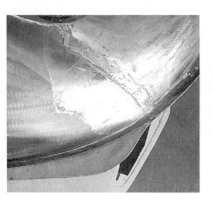

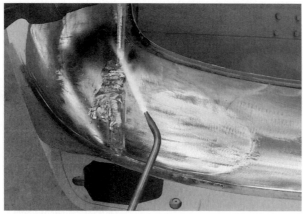

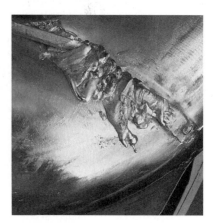

4. Next, the tinning paste is heated until it shines, then wiped over the area to be leaded using a cotton rag.

5. Heat the stick of lead at its base and let it slump on. Don't push it, and don't heat it so much that it runs off.

6. Just let the lead slump into place. Don't try to smooth it at this stage. Put enough on to do the job though.

Leading isn't a dark art but it does take practice, even though pros like Bruce Haye make it look easy. Practicing can be a little financially painful when you consider the cost of lead sticks, but it will be less costly in the long run if you at least have a good idea of what to do before attacking your classic.

To begin with (no matter what type of filler you use) the metal to which it is applied must be absolutely clean and paint- and rust-free. Use a wire wheel and high-speed grinder to get down into grooves, pits and pores in the metal, then use an orbital sander and 80-grit sandpaper to clean the metal further and give it some tooth so the lead will stick.

Next, apply tinning paste with a brush. Just slop on a thick coat over the area to be leaded. This paste has ground up lead and tin in it (40% tin–60% lead) so the container in which it is packaged feels quite heavy. This mixture is also very toxic, so be careful handling it, and wear leather gloves and a protective respirator mask while working with it.

Light up your torch and set it to a soft flame (not too much oxygen) if you are using oxyacetylene. The most common mistake beginners make—and one most of us can ill afford—is applying too much heat. If the flame is too hot, the tinning paste and lead will just melt and run off on the floor. And when you consider the price of lead well . . . you get the picture.

Make sure you put on your leather welding gloves for the next step, then find a clean cotton rag—not a synthetic one. Pass the torch slowly back and forth across the tinning paste. When it starts to become shiny, wipe across it with the rag. Keep working until the entire area to be leaded has a shiny coating of tin. Make sure there are no

7. Dip your paddle in motor oil to help spread the solder. Any type of clean engine oil will do.

8. Heat the lead and the surface then spread the lead with the paddle. You will only get in a couple of swipes before you have to apply more heat.

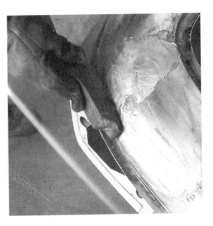

9. Keep heating and spreading the lead until you have it roughly to the shape you want.

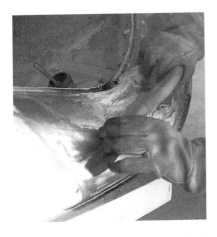

10. Use a fine vixen file to do the final shaping. Work in a crisscross fashion to avoid grooves.

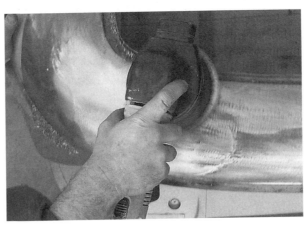

11. An orbital sander and 220-grit disks are the easy way to do the final sanding.

12. The final result looks factory original and will hold up for many years to come.

untinned areas or small pinholes where contamination still exists. If you find any, clean the area and re-tin it.

Next, grab a stick of lead. Heat the surface to be leaded again and keep it warm, but don't get carried away. Heat the tip of the lead stick and let it pile on to the tinned body panel. Don't push the lead into the panel because it will go on too cold and not bond well. On the other hand, don't heat the lead so much that it becomes totally liquid and runs off.

As the lead slumps on to the panel it will pile up and look very rough. Don't try to smooth it by heating it. Just keep melting on more lead until you have enough deposited in place to do the panel. If the lead starts to melt and run, pull the torch further from the surface so you get a cooler flame.

Place a little clean motor oil in a dish or plastic tray, and dip the spoon area of a wooden paddle in it. Heat the lead until it slumps and becomes semi-

molten again, and spread it into place with the paddle. You will only get in a couple of quick swipes before you will need to heat the lead and surrounding metal again. Keep heating and pushing the lead into place until you have it smoothed roughly to the contour you want.

Don't try to hurry this process, and remember, if you use too much heat, everything winds up on the floor. The lead doesn't need to be perfectly smooth at this point. It just needs to be shaped to where it can be cleaned up with a fine vixen file. (A coarse vixen file will take too much off at a whack.) Use the file to smooth the roughness out and clean up the lead until you have it roughly to shape. No heat is required for the filing.

The shaping with the vixen file is the equivalent of the stage where you use a cheese grater file to shape plastic filler. Don't put too much pressure on the file, and work in a crisscross fashion to avoid making grooves and low spots. Be sure to wear at

Use Plastic Filler
- You only need to fill shallow low spots and fix surface irregularities no more than 1/8" deep.
- There isn't likely to be a lot of flexing in the finished panel when the car is put back into service.
- You need to work quickly or are on a strict budget.

Lead Is Required If:
- The panel will be flexed during service or will be subject to a lot of vibration.
- You need to create or fill a damaged edge such as around a door opening or trunk lid.
- You want the very best job possible on an expensive classic. (Just remember that even lead does not last forever. It does deteriorate over time and generally has to be replaced when a car is being restored.)
- You need to fill and shape a low area that will be deeper than 1/8".

13. After master craftsman Bruce Haye works his magic, this car looks like it has always been a roadster.

least a particle mask at this stage, and sweep up all the filing dust and dispose of it properly so people and pets won't inadvertently ingest any of it.

Finish the job by putting a 220 grit sandpaper disk on your orbital sander and going over the entire area until it is exactly the way you want it. Finally, wash the entire area down with a citrus-based cleaner to neutralize any tinning acid left on the metal.

Now you have a repair that will last for many years, and take more flexing and vibration than even the best plastic filler can take. A good working knowledge of both leading and plastic filler techniques will go a long way toward helping you create that show-winning paint job to which you've always aspired.

Chapter 16
Fiberglass Repair

Corvette fenders fracture rather than deform, but the fix is similar to that of a steel car.

Things You'll Need
- Replacement fender
- Plastic panel adhesive
- Pry bar
- Medium weight hammer
- Sharp putty knife
- Angle grinder or die grinder and 36-grit dry sandpaper disks
- Tack rags
- Evercoat Plastik Works SMC
- Teflon palette and a plastic spreader
- Reach-arounds and clamps
- Paint sticks and tape
- Evercoat Rage Gold body filler

First Glass Repair

Corvette bodies don't dent, they crack. The fiberglass panels on America's legendary sports car will take a lot of abuse and bounce back with no ill effects, but if struck hard enough in a collision they will tear and fracture. Fixing such damage is a rather specialized art, and I had only a vague idea as to how to go about it until recently. And then Tom Horvath—my good friend, and coauthor—got a '78 Corvette 25th Anniversary Edition with a fractured front fender into his shop.

Horvath got his start 25 years ago working for a Corvette restorer, so he knows those cars inside and out. Tom says that the only way to fix a broken 'Vette fender is to remove it and replace it, because patches will shrink and look awful in a very short time. Luckily the basic task is not much different than replacing a steel panel except there is no welding involved. Here's how to go about it:

Corvette fenders are attached to the body using bonding strips and panel adhesive. They are also glued to the cowl and inner fender wells with the same adhesive. There are no screws or other fasteners used in their body's construction. So the first thing to do is to pop the old panel loose from the bonding strips and bulkheads. To do this you will need to use a pry bar, available from hardware stores, and a medium weight hammer.

Begin by unbolting the fender wherever it attaches to the frame and body supports front and back. Take out the side marker light as well, and put it where it won't get damaged. Next, gently pop off any trim, such as the badge on our Anniversary Edition, using a sharp putty knife. You can do this because such items are just glued into place on later cars. Now get under the fender and separate the fender well from the inside of the fender using the straight end of the pry bar and the hammer.

Your Big Break—If you sight down the tops of the fenders of a Corvette, you will see a subtle, flat band following the crown of the fender and about three inches down from it. The reason you can see this flat band is because that is where the two parts of the fender are bonded together using a two-inch wide bonding strip of fiberglass. Start at the rear of the strip tapping on the angled end of the pry bar to crack the two halves of the fender apart down the center of the flat band. You may have to hit the seam pretty hard in order to get it to start fracturing.

Carefully pop off trim items using a putty knife. It is safe to do because they are only glued on.

Grind the paint off of the bonding strip that holds the outer part of the fender to its top, then use a pry bar and hammer to crack the seam.

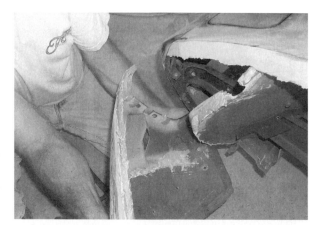

You will need to break the fender loose from the inner fender as well. Use the angled part of the pry bar for that.

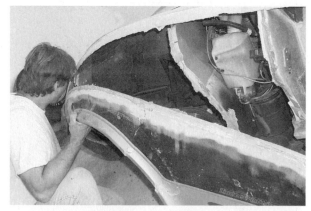

Gently lift up and pop the damaged fender off but be careful not to damage the upper part of the body.

After grinding away the old glue, carefully position and fit your new fender, making sure lines and highlights are straight and the door outline is right.

Tap directly along near the edge of the part of the fender you will be replacing as well. Then get in under the fender and pop it loose from the cowl. Once the fender is detached from the inner structure and the seam has been cracked down the middle, gently lift up on the bad fender from the bottom and keep tapping until it pops loose. Don't become impatient though, because you don't want to damage the part of fender above the bonding strip.

Stripped Naked—Next, carefully crack the bonding strip from the damaged fender. You will want to reuse this item, so try not to break it. Next, use an angle grinder or die grinder and a 36-grit dry sandpaper disk to remove the old adhesive from the cowl and fender well. (In our case, Tom had to replace the inner fender well, but you may not need to.) Any glue left on the bulkheads will cause a bump or raised place when the fender is glued on, so be sure to get the old adhesive off completely.

Pick up your replacement fender panel from Chevrolet. The new panel will be a gray, injection molded piece of plastic rather than the old-style laid up panel with glass cloth on the underside. These panels don't look tough, but they are. Wipe the entire fender down with a good degreaser, and don't handle it with dirty hands because grease and oil will prevent the adhesive used to install the fender from sticking.

When that is completed, rough up the new fender panel on the inside where it is to be glued to the cowl and bulkheads, and make sure you do the area where the bonding strip attaches as well. Use a compressed air nozzle or tack rags to get rid of any dust that might have accumulated on the car or the new panel before attempting to bond parts together.

Rough up the new fender and other attaching flanges with a grinder and 36-grit sanding disks.

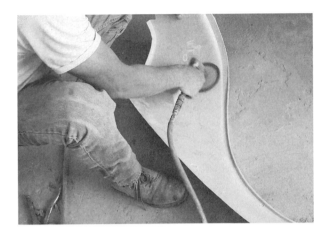

Tom uses Evercoat Plastik Works SMC Panel Adhesive to attach the bonding strip and the fender.

Mix the adhesive the same way you would plastic filler. A golf ball size gob with an inch of catalyst is the right amount to work with.

Reattach the bonding strip to the upper fender using the adhesive. Be sure to butter both components before pressing them together.

Attach the fender on the car using clamps and reach-arounds making very sure the fit is exactly what you want.

Together Again—Tom uses Evercoat Plastik Works SMC panel adhesive for the next step. This material is a lot like plastic filler but it cures much harder. Mix a golf-ball-size gob of filler together with an inch of catalyst using a Teflon palette and a plastic spreader. Work in a figure-eight pattern and don't lift the spreader because you could fold in bubbles, which will weaken the adhesive. Mix the two components together until the color is consistent, with no streaks in it.

On a warm day the adhesive will set up in about 15 or 20 minutes, so you need to work fast. Attach the bonding strip to the new fender by buttering both sides and pressing it into place. Wipe away any excess adhesive from the bonding strip flange. Now let the bonding strip set up for a few hours before going further.

Fitting—Fit the fender carefully using reach-arounds and clamps to hold it in place. Make sure the highlight lines and the bottom of the fender are not cockeyed, and make sure the fender makes a tight fit at the integrated bumper. In all likelihood your fender will be a little long, so it will come back too far into the door gap.

This is done intentionally by the manufacturer so you can shape the fender to the door. Use a die grinder to get the edge of the fender roughly to shape, but be careful. You can take material away, but you can't add it back if you over do it. Finish fitting the door gap by hand using 36-grit dry sandpaper. Carefully fitting and positioning the fender before gluing it is crucial. Otherwise your 'Vette will look forever funny.

Take your reach-arounds off only as needed to do your gluing. Apply adhesive to the cowl and inner

Glue the fender to the fenderwell and cowl first, then force glue in along the bonding strip. Make sure the adhesive penetrates well, and fill any gap or recesses with glue.

Bolt the fender in place at the bottom and in front and then attach a paint stick to the fender about half way down using wide masking tape.

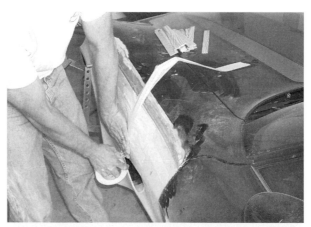

Tape a second stick so its top is about halfway down on the first stick, then bend it over and tape it.

Bend the stick over to apply pressure to the upper fender, and tape it securely to the hood.

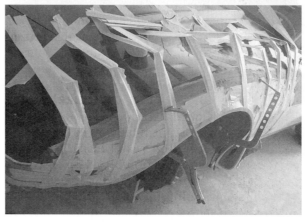

Keep taping sticks about 8 inches apart until the whole fender is under even clamping pressure along its seam.

fender first and be sure to butter the fender where it attaches as well. Work quickly because the half life of the adhesive is only about 15 minutes on a hot day. Don't be tempted to fiddle with the ratio of plastic to catalyst because you could weaken the bond if you do.

Finally, loosen the fender from the grip of the reach-arounds only where you are working, and gently push up on it from underneath to allow you to force adhesive between the flange and the bonding strip. Make sure the glue penetrates well, and make sure the fender doesn't get cocked or misaligned while you work. Once it's on there and set up there is little you can do to correct any problems.

Stick Around—If you pick up nothing else from this chapter, here is a tip that is invaluable when working on Corvettes: Use paint sticks and tape to clamp the fender into place instead of drilling holes

and installing small screws or pop rivets. Tom showed me this trick and I was duly impressed. He first bolted the fender in place where it attaches at the bottom and in front and then he got out the paint sticks and tape.

You see, the usual method of drilling holes and installing flush mounted screws—even though you take the screws out later and fill the holes—can bruise the fiberglass and weaken it. Also, fiberglass is organic and flexible unlike metal, so the plastic around the screw holes will shrink and cause dimples that will soon show up in the car's finish. Also, rivets left installed will work loose due to

Let the fender cure overnight, then grind away the excess glue using a die grinder.

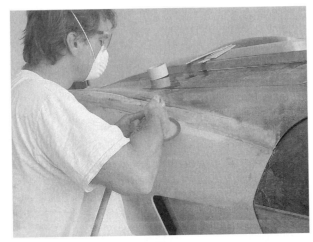

Use the same plastic filler you would on a steel car and sand and finish it the way you would a steel fender.

Taking paint off of a fiberglass car is a special challenge. One way is to use a sander and 300 grit open coat sandpaper.

flexing and little holes will crack out in the finish.

Use wide masking tape to attach the paint sticks, and tape the first one securely to the fender about half way from the top of the fender. Don't be stingy with the tape because it will be under stress to come loose when you bend the stick over. Gently bend the paint stick over to create pressure on the fender and then tape it securely to the hood.

Tape a second paint stick so its top is about half way down the first one. Now bend it over and tape it to the hood as well. Use enough tape to prevent the sticks from working loose. Keep working along your fender, spacing the paint stick clamps about 8 inches apart until you get to the front of the fender. To finish up, go back and double tape the sticks down making sure they apply approximately equal tension from one to the other. Check your work carefully to make sure everything is properly aligned and fitted, then let the glue cure over night.

When the fender mend has fully cured, use a grinder and #36 sanding disks to grind away any adhesive bumps and irregularities, and to create a tooth (slightly roughened surface) for plastic filler. You can use the same plastic filler on Corvette fiberglass fenders as you would on a steel car. The same rules apply. Tom prefers Evercoat Rage Gold but there are other good brands as well.

Mix a golf-ball-sized gob with an inch of catalyst and apply a layer no heavier than an eighth of an inch thick, with the filler at its thickest down the center seam and feathered out on either side. The final steps are the same as you would use to finish a repair on a steel bodied car. Sand the filler to the correct profile, use glazing putty to fix small irregularities, then block sand. Check your work using a guide coat and more sanding before priming and painting.

First Glass Paintwork

Recently I had the chance to pick up some paint and finish tips from Edwin Alvarez at J&D Corvette in Bellflower California where they do first-rate professional restorations of classic Corvettes. When they get through with a car its paint looks so deep and dazzling that you could almost go swimming in it. There is no hint of grooves, flat spots, strips, dimples, or other blemishes anywhere in the finish.

To achieve such a level of perfection takes hours of preparation before the car is painted, and more hours color sanding before it is done, but it is worth it as you can see. None of it is easy, and it's anything but quick, but the results are spectacular. Of course the final finishing process is color sanding, and that is what makes the difference between a trophy winner and an also-ran, but prep on a fiberglass car is especially important. Here's how it's done:

Alvarez tells me that painting a plastic car such as a Corvette is not like painting a metal-bodied car. That's because fiberglass moves, shrinks and changes with time. Even the most perfect fiberglass body will in time start to show the seams where it was glued together, and will show small, shrunken flat spots as well.

We won't go into the actual shooting of the paint because that is essentially the same as when painting a metal car, and is covered in another chapter, as is color sanding, but we will talk about some of the unique problems fiberglass car painters face.

Stripping Paint—The major challenge that faces Corvette restorers is getting the old paint off. You don't want to use conventional methyl chloride paint strippers because they can cause damage to the plastic below. About the best way to remove the old finish is to use an air or electric circular sander and 300-grit open coat dry sandpaper. This process, using such fine sandpaper takes more time, but it also minimizes the risk of damage to the fiberglass underneath.

Another advantage to sanding the old finish off of your 'Vette is that doing so also takes out the little flat spots and shrunken areas that have developed in the body surface, and it helps clean up and hide body seams. Just be very careful not to linger on sharp fender crowns and edges while sanding. In fact, if you want to strip your Corvette, find a damaged fiberglass panel to practice sanding on, because if you aren't confident and experienced with a spinning sander you could do some real damage. Small problems can be fixed with a thin layer of filler later, but deep grooves would be an expensive mistake.

And don't get impatient. Coarser sandpaper would take the paint off faster, but it could leave deep grooves that would weaken the car's body. And given that fiberglass panels are more flexible than metal ones, and that the panel is weakened, any filler you used to hide deep mistakes would be very liable to cracking. Plastic filler works well if it is not bent or subjected to vibration, but it was never intended for highly flexible panels.

Before beginning to strip the car of old paint, remove all of the trim and triple tape anything you might accidentally scratch, such as window glass.

Here is the final result after painting; dazzling and ready to take home a trophy.

Put on at least a particle mask before starting, and then begin sanding. Do not apply heavy pressure, and keep the sander moving at a steady pace at all times.

Let the sandpaper do its work, and when it starts to clog, stop the sander and brush the dust out of the disk. As soon as the sandpaper starts to lose its 'tooth' or cutting ability, change the paper rather than increasing the pressure. Otherwise you risk damage to the car's body. Just remember, sandpaper is cheap, but body panels are expensive.

Shooting Paint—Painting a plastic car is essentially the same as painting any other car, and exactly how you go about it depends on the paint system you choose. Suffice it to say you should always stick with just one brand of paint and one paint system because as we have said in other chapters, modern paints are very sophisticated chemistry and may not be at all compatible with one another. Most cars these days are painted with the new urethanes because they are so durable, and because they help meet environmental rules concerning volatile organic compounds in the atmosphere, so that is what we will be working with here.

Make sure you shoot on enough paint to allow for color sanding. One-stage urethanes must go on a little thicker to allow for sanding, and with base-coat clear-coat systems, you should shoot on a little extra clear coat to allow for color sanding too.

Chapter 17

Spray Guns, Compressors & Spray Booths

Paint guns come in these two styles. On the left is a gravity-feed type, and on the right is the old-style suction-feed gun. They also come in high velocity, low pressure (HVLP) for environmentally sensitive localities as well as the traditional high-pressure-type guns for the few localities where they can still be used. And if you are going to be shooting waterborne color coats, you will want a gun made of stainless steel to avoid rust.

Spray Guns

If you don't count the spray booth, there is actually very little equipment required to shoot paint on a car. All you need is a source of clean, dry, compressed air and a quality paint gun—albeit one that has been carefully maintained. Perhaps that's why professional painters—no matter where they work—use and maintain their own personal paint guns almost exclusively.

In fact it's definitely not good manners to use another painter's gun unless he authorizes you to. That's because some kinds of paints, if left in the gun longer than half an hour, will make the paint gun virtually useless. And guns used for shooting waterborne paint must be cleaned extra carefully. So if you want to get serious about painting cars, I suggest you get the best spray gun you can afford, and maintain it just as carefully as the pros do.

There are plenty of cheap knock-offs of professional paint guns available at swap meets for bargain prices, but you don't want them except perhaps for shooting primer. They are poorly made and poorly machined and will not do a good job even in experienced hands. If you are a novice painter, give yourself every break possible because there is a

learning curve to shooting paint properly and you won't want to make it steeper for yourself. Buy a decent paint gun to eliminate as many problems as possible up front.

There are a number of good brands of paint guns—among them Sata, Sharpe, and Binks—but like a lot of long-time professional painters, I use DeVilbiss exclusively. I have a conventional, high-pressure DeVilbiss paint gun that I bought 30 years ago and it is still as good as the first time I used it. I can't legally shoot paint with it anymore in California, but I still won't part with it.

Some folks are Chev people and some are Ford people. I'm a DeVilbiss man, as was my father before me. The DeVilbiss company got started in the 1890s making spray equipment for perfumes and water, and when automotive spray painting got going in the early '20s DeVilbiss made some of the first equipment. Lots of improvements have been made in paint gun technology since those days, but the company's quality has remained tops.

Suction or Gravity?—I recently acquired a new DeVilbiss gravity-feed gun with the paint cup on top that I really like. It is lightweight, simple, and because the cup is on top I don't run the risk of bumping it against my work when

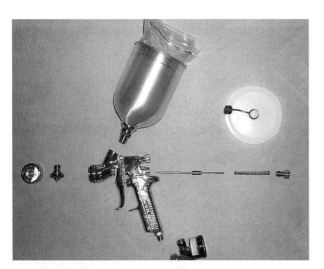

Here are the parts of a typical paint gun.

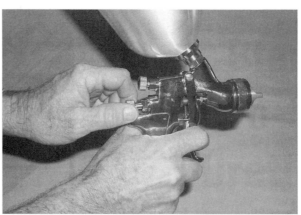

Lower knob adjusts the atomization air and material flow.

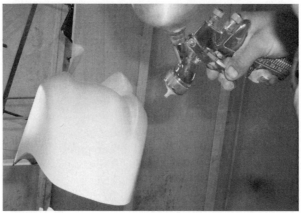

Hold an HVLP gun about two to four inches from the object being painted. With a conventional high-pressure gun, hold it about eight inches away at all times. waterborne paint guns need even more room at about 12 inches from the surface.

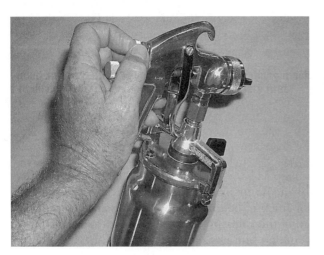

Upper knob adjusts the pattern width.

shooting tops. It is also much lighter than the old-style suction guns too, and that makes it easier on my shoulder.

Of course, the old-style suction-type guns with the cups on the bottom have their advantages too. For example, you can get down lower with a suction gun because when you tip back a gravity feed gun with the cup on top to paint a rocker panel, the paint inside can end up tipped at such an angle that it won't flow into the gun. Of course, if you can afford both, get one of each. And if you really want to go first class, pick up a third cheap gun for shooting primer, because some primers are corrosive and can mess up a good gun over time.

More recently, fittings have become available to allow disposable paint cups for gravity feed guns. These are especially nice because the mixing measurements are clearly marked on the side of the cup, allowing you to mix paint right in them. And

then they can be disposed of easily when you are through shooting. (Let the vestiges of paint dry inside the disposable cup before tossing it out. It is against the law in many areas to dispose of liquid paint in household rubbish.) With disposables, there is no more sopping paint out of the bottom of the cup and cleaning the air passages with toothpicks and pipe cleaners.

The Advent of HVLP—In the last thirty years there has been a big push on to cut air pollution, and as a result a new kind of spray equipment called HVLP (High Velocity Low Pressure) has been developed. These alternative paint guns look the same as the older types to the untrained eye, but the spray orifices in the caps are much larger. HVLP equipment comes in both suction and gravity feed configurations, and touch-up guns are also available. And the new HVLP equipment is no more difficult to use than the old-style high-pressure equipment.

Conventional high-pressure spray guns are still legal in some localities, and there are painters who still prefer them, but most of us are switching over to HVLP paint guns even when we are not required to do so by law. That's because whether or not you care about damaging the ozone layer or global warming, you will most likely care that an HVLP paint gun puts at least one third more of that $200–$400 a gallon paint on your car, and consequently wastes one third less of it in overspray.

The old, high-pressure guns that were standard in this country for 60 years worked on the Bernoulli principle. They used air rushing at 35–65 lb pressure past a tube that went down into the paint cup, thus creating a vacuum to suck the paint out and atomize it. The new HVLP guns also use air rushing past a tube to pull the paint out of the cup, but at only about 10 lb pressure. And the guns shooting waterborne color coats are only set at about 6–7 lb of pressure, but this may vary according to your technique.

If you haven't tried an HVLP paint gun yet, you will be surprised at how easy they are to use. I had no trouble switching over, and I sure don't miss all that sticky overspray in the air while painting. I also like the fact that there are much less VOCs (Volatile Organic Compounds) wafting around to cause an explosion hazard and pollute the atmosphere. But most of all, I like the fact that I can paint a car using much less of that paint that costs nearly as much as my wife's perfume these days.

Spray Technique—Learning to shoot paint properly takes a little practice, but it is not hard to do. If the item to be painted is prepared properly, the paint is mixed according to the instructions, and the ambient temperature in the spray booth is between 65 and 75 degrees (85 degrees is preferable for waterborne paints), everything should go just fine. Provided you've taken the time to practice, that is. Get some old metal objects such as crumpled fenders, coffee cans and barbecue hoods and practice on them before trying to paint your car. Here's how to get started:

Begin by installing a liner or cup if you have one of the new paint guns that takes disposable cups or plastic liners. Mix your paint to the exact proportions stated in the directions, and then strain it through a 60 or 90 mesh filter as you pour it into the cup. Most painters don't fill the paint cup to the top because it will be more prone to leaking that way, and it makes the gun too heavy for easy management.

Before attempting to paint anything, adjust your air pressure and spray pattern by spraying a piece of paper hanging on the wall. Use a small valve and gauge attached to the bottom of the gun—or one that attaches to your belt—to set the air pressure at the recommended figure for the type of paint you will be shooting in the case of an old style gun. If you are using an HVLP paint gun, set the inlet pressure at 30 psi, which will give you 10 psi at the cap, which is all the law allows.

With an HVLP gun, open the spreader valve all the way. With a traditional gun, adjust it to produce the right pressure for the paint you are going to be shooting. Then use the lower knob, which is the needle adjusting screw, to obtain a full wet coat. When you have the pattern the way you want it, begin spraying. Always paint a car in the direction of the air flowing through the booth. In other words, start where the air enters the booth and finish where the air exits the booth. That way you avoid getting your own overspray into your work.

If you are using an old-style high-pressure paint gun you need to hold it about eight inches away from your work. That is about the width of a man's hand from pinky to thumb, fully spread. If you are using an HVLP spray gun, hold it between two to four inches from the work and if you are using a waterborne system 12 inches is the correct distance. Also hold your paint gun exactly at 90 degrees to your work and pivoting your wrist to accommodate changes in contour.

If you tip the gun at an angle, more paint will go on at the part of the spray pattern that is closest to the gun, and the other end of the pattern will be a little dry. This can be disastrous if you are shooting metallics because the mica in the paint won't lie flat in the thicker areas, so streaks will form in the finish. Only let up at the ends of passes with the gun enough to stop the flow of paint but not the flow of air. That's because pressure can build and cause a paint buildup when you first hit the trigger again.

Move in evenly timed passes that put on a full wet coat without sags. With an HVLP gun, if the paint goes on too thick, pull back a little from the work. If it goes on too thin and dry, get a little closer, or slow down a little with your passes. With a high-pressure gun you can also adjust the air with the upper knob on the back of the gun. Move along with your paint gun to keep it at the right angle and distance from your work. Never stand in one place and move your arm in an arc. You'll get dry and wet areas if you do.

Overlap your passes by 30–50 percent. Try to get a rhythm going so you are moving at the same speed and giving each area equal attention. As the paint runs low in the cup, don't try to shoot every last drop because the gun will sputter and be inconsistent, making a mess of your work. If you get any sags, leave them alone. Trying to wipe them out with a rag will only compound the problem.

Usually, if the paint sags though it has been mixed correctly and you are holding the gun at the correct distance, it can be corrected by decreasing the atomization air pressure using the fluid needle adjustment screw (lower knob). The pattern width

Spray Pattern Problems

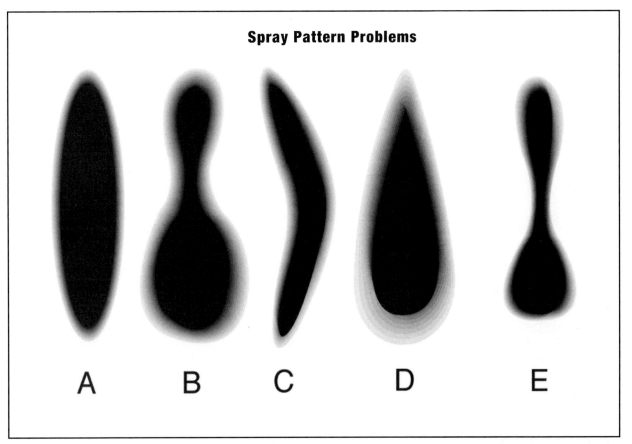

A   B   C   D   E

A: The way a spray pattern is supposed to look. B: What happens when the air pressure is too high, the fluid flow too low, or the spreader adjustment valve set too high. C: This happens when the left- or right-side horn holes are plugged. D: Material flow is exceeding air cap's capacity, or the spreader adjustment valve is set too low. E: The horn holes are plugged, or there is an obstruction on the top or bottom of the fluid tip.

can be adjusted by turning the upper, spreader adjustment valve.

Keep your air hose out of your work by attaching it to your belt with a clip, throwing it over your shoulder, or even running it down your trouser leg. If your spray pattern becomes skewed or the gun starts to sputter, stop and determine what is wrong. Common problems are outlined above.

Whether you paint a whole car all at once or paint it in pieces is up to you. There are those who claim that painting a complete car all at once provides a more consistent finish, but I like to take the car apart as much as possible so I can easily get up into fenders and painted areas that cannot be reached when the car is assembled. It all depends on how much time you want to spend and what the car really needs.

If you paint your car all together, start by painting door jambs and hard to get areas first, then go to the sequences shown in the accompanying illustrations depending on the type of spray booth you are using. Most painters like to shoot on a light misty tack coat first, let it get sticky, then shoot on full, wet coats. How long you let the tack coat set is a matter of choice. Tom Horvath likes the tack-and-whack method where you only leave the tack coat for about five minutes. Bruce Haye gives the tack coat 20 minutes to set up.

I have included two chapters outlining the way these two top professional painters do it. Read them both, then take what you need from what they have to say and develop your own technique. As we have said before in this book, painting cars is an art. There are several good ways to do it, but practice is what makes the pros.

Cleaning Your Paint Gun—As we pointed out in the paint chapter, the newest finishes are essentially plastics. They harden in a given amount of time by chemical reaction (depending on the temperature) and that's that. Once they set up, they can't be dissolved with the usual thinners. In fact, the new polyester primers are so tough and solvent resistant that if you leave them in your paint gun for longer than 25 minutes you might as well throw the gun away.

As a result, cleaning a paint gun isn't just tidying up the outside of it. To do the job right the gun must be torn down and thoroughly scrubbed with gun wash or lacquer thinner. Automotive paint dealers as well as the people who manufactured your spray gun can sell you a brush set for cleaning your paint gun if the equipment didn't come with a the gun set already. In addition to this set of stiff bottle brushes of various sizes, I also keep a supply of pipe cleaners on hand, as well as a box of wooden toothpicks.

When you are through painting, take your paint gun apart and clean it with lacquer thinner or gun wash.

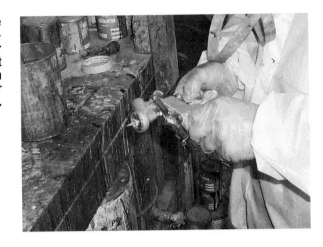

Clean the orifices in the air cap with toothpicks soaked in lacquer thinner. Never use wire for this purpose because it can damage the air cap.

Here is a typical paint-gun cleaning kit. The pipe cleaners and toothpicks are my idea.

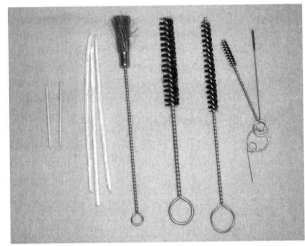

The cup is unscrewed, the air cap pulled off, and the needle valve removed as part of cleaning.

My new DeVilbiss gravity-feed gun uses lift-out disposable plastic liners, so cleaning the cup is easy, but the older guns can be messy. First you must pour any excess paint into a can and seal it. Then you need to slosh a fair amount of lacquer thinner into the cup and wipe it out with a rag. Make sure there isn't even a hint of paint left in the cup, then clean the siphon tube using a brush.

Put a little lacquer thinner in the cup, hook up some air and shoot it through the gun until the

thinner comes out clear. Put your finger over the tip of the gun and spray again to back-flush the system, then disconnect the air to the gun. Remove the air cap retaining ring and air cap. Drop these into about one inch of lacquer thinner in your paint cup and let them soak while you remove the fluid needle. Do this by backing out the lower knob on the back of the gun. You may need to depress the trigger to get the needle to slip back out of its packing.

Wash the needle in lacquer thinner, then put a little paint-gun lubricant on the area of the needle where the packing gland goes, and at the trigger pivot points as well. Don't use ordinary grease or Vaseline for this because it could get into the paint and cause fisheyes, especially with waterborne paints. Use pipe cleaners and small brushes to clean all tubes or orifices depending on how your gun is constructed. Never clean the air cap ports or needle orifice with wire or metal objects. These openings are precisely machined and can become damaged if you do. Instead, use wooden toothpicks dipped in lacquer thinner.

There are some painters who immerse their entire paint gun in solvent to clean it, but this has drawbacks. When you do that you wash all of the lubricant out of critical areas of the gun. Instead, take the time to clean the gun the right way as outlined above each time you use it, and don't ever put off cleaning it until later.

Guns that are put away that have not been properly cleaned can sometimes be saved depending on what you were shooting in them, but if you have to take the gun to an expert at the paint supplier, it may cost you almost as much as the gun is worth to save it. A clean paint gun will do a good job, and a dirty one can cause a disaster that might even mean stripping off the new finish and starting over, so keep your paint gun scrupulously clean at all times.

Compressors

If you are serious about paint- and bodywork, you'll need an air compressor. The bigger the better. And yes, it is possible to paint a car without a compressor with the new HVLP systems, and if you are a home hobbyist who is only going to be painting occasionally you may want to go with one of those self-contained systems made by Accuspray. I have one and I like it for small stuff, but these days, I have a two-cylinder, two-stage monster compressor with a tank big enough to bathe in.

With a compressor you can drive sanders, impact wrenches and buffers and you can media blast paint and rust from old car parts in minutes. You can get by with a small compressor if all you want to do is paint, but I would suggest you buy the biggest compressor you can afford if you are serious about body- and paintwork on cars because in addition to painting, you can operate so many useful power tools with air.

Two-Stage Models—Get a two-stage model if possible. You can tell if a compressor is two-stage because it will be at least a two cylinder model, and one cylinder will be bigger than the other. Also, for media blasting, your compressor should have a big tank. You want a tank with at least 60–80 gallon capacity if possible. Media blasting takes lots of air, and even a big compressor with a small tank will run continuously, get too hot, and wear out quickly.

Go for a five horsepower model if possible. It is hard to do media blasting with a one to four horsepower compressor, and media blasting is great for cleaning up smaller items. Also, unless you are working in a building without power, get a compressor that has an electric motor rather than a gasoline engine. The gas-powered models do have the advantage of portability, but they require double the horsepower to produce the same cfm (cubic feet per minute) of air.

You will also need at least one (two are better) water trap, because water condenses out of compressed air and will get into your blasting media or into any paint you are spraying. Water traps should be mounted low in the system so as to take advantage of gravity to help drain the water.

Single-Stage Models—Inexpensive compressors are generally single-stage types. Such compressors (depending on size) can produce plenty of air, but because of their design, the air they produce will contain a lot of water. This is due to the heat generated by high pump speeds. And because of these high pump speeds, single-stage compressors don't last as long as two-stage types either.

Another problem with one-stage compressors is that they are not capable of producing high

Get the biggest compressor you can afford. This is Tom Horvath's at Tom's Custom Auto Body.

Buy a two-stage compressor like this if you can possibly afford one because it will last longer and the air it produces will be less moisture laden.

The Accuspray is a convenient, inexpensive, portable, and an excellent choice for the hobbyist who will only paint occasionally and doesn't want to invest in a compressor. It's available from the Eastwood Company.

pressures (psi) in the storage tank, so they must run more often to keep the tank topped up. The higher the pressure that can be stored in the tank, the more air there is to use before the compressor has to come on to pump it up again.

Two-stage compressors have two cylinders of different bores. Single-stage compressors are often single cylinder, but may have more cylinders of the same bore. The way a two-stage compressor works

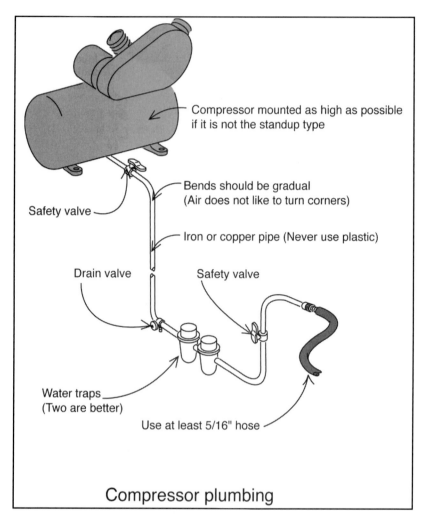

Compressor mounted as high as possible if it is not the standup type

Bends should be gradual
(Air does not like to turn corners)

Safety valve

Iron or copper pipe (Never use plastic)

Drain valve

Safety valve

Water traps
(Two are better)

Use at least 5/16" hose

Compressor plumbing

Mount the compressor up high if possible, use gentle bends in the piping and add shutoff and safety valves along with water traps.

is, the bigger cylinder compresses the air to an intermediate level, then the smaller cylinder compresses it further. An intercooler is used to cool and dry the air.

Two-stage compressors run more slowly and quietly and produce cooler, drier air. They are also designed to develop pressures of 160–200 psi. A two-stage compressor with a 60–80 gallon tank is the ideal choice for media blasting and power tools, but if you can't afford such equipment, you can still do most chores with a more modest compressor.

The Accuspray—This is not a tool for the pros, but it is excellent for the hobbyist on a budget. It needs no water traps and has its own source of air. The spray gun works like any other HVLP type gun, and is good quality. Shooting paint with one of these is no different than the old way using a full compressor. The portable four-stage compressor is maintenance-free except for the air filter on its side, which must be replaced now and again. I have an older three-stage version of this tool at my shop and

I use it to paint smaller items and subassemblies. It has been a real time-saver. I got mine from the Eastwood Company.

Compressor Specs—Most people think psi (pound per square inch) is the best measure of a compressor's capabilities. Not so. The most important figure is the cfm, or cubic-feet-per-minute it can put out. Air compressors are rated at cfm at a given psi. A compressor that produces 17 cfm at 100 psi can store a lot more air than one that produces 17 cfm at 60 psi. So the best compressor is the one that produces the highest cfm at the highest psi.

The cfm rating of a compressor is figured the same way as engine displacement. Manufacturers measure the bore and stroke of the cylinders, then work out the swept volume and use that figure in their literature. However, compressors—like automobile engines—have less than perfect volumetric efficiency. Generally, the actual efficiency is likely to be around 50 percent of the stated figure for a cheapy, to 70 percent for a quality compressor. Therefore, assuming you buy a good 16 cfm rated unit, it will likely only give you about 11 cfm.

Sears Craftsman, heavy-duty, two-stage air compressors sell for between $500 and $1,200 and are backed by their famous warranty. For single-stage compressors, Campbell Hausfeld and Ingersoll Rand are respected names. Single-stage compressors cost between $300 and $500.

Spray Booths

When I was in my early twenties I remember helping a friend paint a car in his driveway. In those days it wasn't total folly to do such a thing because the lacquer we were using dried almost instantly if you mixed it with a fast thinner. As it was, we chose a hot day to do the deed and the thinner was actually too fast, so the paint started drying in the air and looked more like powder on the car. We were only saved by the fact that you could always sand lacquer and recoat until you had a finish you were happy with.

As forgiving as lacquer is though, we still wound up with a marginal paint job because bugs, dust, dandruff and other airborne torments beset us from the start. We didn't realize that the car, the driveway and the very air we were breathing were full of contaminants, and that our car had a slight static charge caused by the dry day and gusty wind, and that drew in dirt and small creatures like a magnet.

We also didn't realize that it wasn't a good idea to paint a car in the same spot where you had sanded it. We wet down the driveway before shooting the paint, but it was a sunny day so it didn't stay wet

Without continuous air circulation, fumes can build up and cause a fire, and your partially cured overspray will wind up on your work.

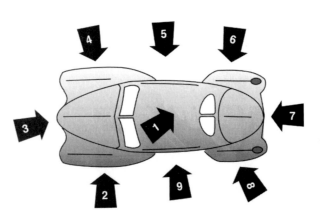

Here is the order in which you will want to paint a car depending on which type of booth you will be using.

Order for cross-draft booths

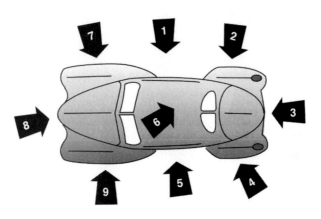

Order for down-draft booths

long. And wetting down the drive had its own unpleasant side effects because a few drops landed on the car's rocker panels and they didn't evaporate.

In the end the car didn't look that bad from 10 feet away, as they say, and I guess we saved some money. But these days, the contemporary urethanes just can't be sprayed in such environments, and you can't really do a cheap paint job at home for less than the bargain shops. Besides, it is illegal to paint outdoors in most localities, and your paint job would turn out terrible at best.

The only sensible way to paint a car now is to do it in a spray booth. And in most localities, if you don't have a professional paint booth, you will need to rent one. In larger metropolitan areas there are rent-a-booth companies that make their money by renting out paint booths and equipment and that is a good way to go if you don't have access to such faciities. Another possibility is to buy time in a spray booth from a body shop. Still another is to enroll in a class at a local trade school.

Types of Booths—There are several types of spray booths. There is a flow-through type where the air is pumped at one end and goes out through massive filters at the other end. Another type of booth is the down-draft type where air is pumped in through filters in the ceiling and exhausts down low along the sides of the booth.

But the best, full zoot professional spray booths are positive atmosphere types. These create a positive pressure inside the booth that keeps contaminants out. The ultimate in this situation is the down-draft booth in which air is fed in from the top and is exhausted out through the floor of the booth. The main thing to keep in mind is that the type of paint booth dictates the order in which you paint a car. The illustration above shows the details.

Whether you rent, beg or borrow a booth or have

your own, be sure it is completely clean before bringing a car into it. If there is any dirt anywhere, wipe the booth down with a damp cloth to remove it. Mop the floor carefully too. And of course never sand a car in the paint booth. Also, check to see that all of the lights are working properly because you will need plenty of light to see what you are doing.

Check the large paint filters to make sure they are not clogged. This is especially important because if air stops flowing it could ruin your paint work, and fumes could build up and cause an explosion. It has happened, and the results can be catastrophic. Also check to make sure the seals at the doors of the booth are intact. If they are cracked or damaged in any way, pack the area with wet towels to keep contaminants out.

Make Your Own Booth—If you live out in the country in a state that still allows it, it is possible to

There are two basic kinds of spray booths— down-draft and cross-draft.

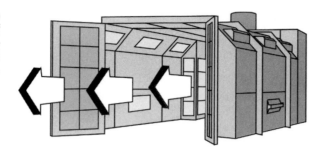

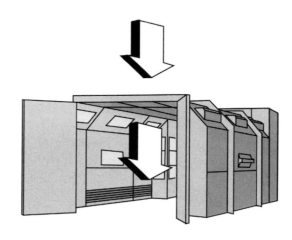

put together your own spray booth, but check with local authorities before you attempt it. The fines for polluting the atmosphere can be very heavy. Also, people have painted their cars in their garages (I did it myself in the old days), but that can be very dangerous.

Water heaters and electrical connections can easily cause a spark or flame that could ignite paint fumes and cause a fire or explosion. If your garage is detached from your house, and you make sure all the electrical connections are explosion-proof, you could get way with shooting paint in it, provided you have proper ventilation. And though, with a fresh air pack such as those made by Survivair you won't actually be breathing the fumes, the buildup

can be dangerous without adequate ventilation, and your own overspray will end up falling back into your work and spoiling it.

Using Your Garage—As I said before, I really don't recommend it for safety reasons, but if you do decide to paint your car in your garage, put large fans with explosion-proof connections at each end, and have them drive the air through large air conditioning filters. Another trick hobbyist painters use is to staple plastic sheets similar to what is used at construction sites to the walls and rafters to contain the dust and contaminants that have lodged on them.

The next challenge is light. You need a lot of it from every angle to paint a car. A dozen or so fluorescent lights mounted on the walls and ceilings will do the job, but again, they need to be hooked up with explosion-proof connections to prevent sparks. Professional paint booths are wired inside steel tubing, and there are special fittings built in to it that stops the spread of fire inside the conduit. If you are going to set up your own booth I suggest you install a few of these at critical junctures too.

A third alternative is to build a sort of greenhouse-like structure out of strips of wood and clear plastic tarps that will allow you to spray your car outdoors on a sunny day. Just keep in mind that you still need plenty of ventilation in any case. Never set up such a booth on dirt or grass though because you will blow all kinds of dirt into your work if you do.

Of course, it is also possible to paint an older car with bolt-on fenders in pieces in a smaller booth. I like to paint cars in pieces myself because it allows me to paint all the panels completely so there is no problem with rust developing down in cracks and crevices. In the end, you will still need some sort of paint booth to paint the main body of the car though.

Chapter 18

Paint Tech

This Pantera is shot with a single-stage urethane. The finish looks deep, wet, and magnificent.

A lot of progress has been made in paint technology in the last few years, though automotive paints can still be divided into two basic categories. There is lacquer, which dries by evaporation, and there is enamel, which dries chemically. Based on those definitions, almost all automotive paints used today are enamels. However, today's urethanes—though technically classed as enamels—have no more than a technical similarity to the old synthetic enamels of the past.

Urethanes are made with acrylic resins and various polymers and are actually a very resilient, tough form of plastic. They are far superior to the finishes of a few years ago, and they are designed to meet the strict environmental laws in many states. To do that, they have been developed to be very high in solids such as pigments and binders, and as low as possible in solvents that pollute the atmosphere.

But paint companies are understandably reluctant to talk about what exactly is in their paints because that might give away trade secrets. On the other hand, it is much easier for the average individual to achieve good results with modern primers and paints, provided the instructions of the manufacturer are followed exactly. Let's see if we can untangle some of the complexities of today's finishes.

To begin with, the paint systems used at the factories are quite different from those used in the refinishing industry. Auto manufacturers now use electrostatically applied primers that bond extremely well, but these require special equipment to apply, and the car must be disassembled at the time. The way it works is an electric current is passed through the parts, that attracts the primer into every crevice and cranny. That also helps the primer to bond. Replacement parts such as fenders and hoods generally come with electrostatic primer on them as well, and this is referred to in the industry as an E-coat.

Factory color coats are quite different too They are most likely waterborne, their chemical composition is designed to cure quickly, and it is designed to be baked on the individual parts. The heat lamps used by aftermarket automotive refinishers are limited by the fact that you can't get an assembled car so hot that its rubber and plastic components melt and its glass cracks.

Paints intended for refinishing go on thicker, take longer to cure, and are chemically different than OEM (Original Equipment Manufacturer) paint. The primers used with these aftermarket paints are different too. But this does not mean that an automotive refinisher's work won't last as long as a factory job. After all, the factory is mainly interested in using as little paint as possible and having it cure as quickly as possible. A quality refinisher can do paint work that far surpasses the factory in beauty, shine and durability.

In fact, one of the common maladies you see today on cars with base-and-clear coat paint jobs is when the clear coat gets too thin from wear and oxidation after a few years so it doesn't shield the paint below it from ultra-violet rays. Then the clear coat delaminates from the base coat.

Base coat and a clear coat of urethane plus color sanding are the secrets behind this deep dazzling paint job.

Here's a base and clear candy paint job. The base is gold metallic, and the clear is tinted green to produce this deep finish.

The same thing happens when single-stage paint gets too thin due to polishing. It begins delaminating from the substrate below and starts flaking off. In fact, once a modern finish gets below 2.5 mils thick, this is likely to happen, and factory paint is only about three mils thick. The factory is only concerned that the original paint last five years anyway, because that's how long most people keep new cars.

A refinisher's work will be in the range of 5–7 mils thick, so delamination isn't a problem. Of course, paint can be too thick at around 8–10 mils, at which point it will be prone to cracking, especially in areas that flex. And thick paint will "move" too because the outer layer will get hotter than the lower layers in the hot sun, so the outer film will be prone to shrinking, crazing and alligatoring no matter what kind of paint you are shooting.

Nitrocellulose Lacquer

Until 1956 there was just nitrocellulose lacquer and synthetic enamel. Lacquer was used for fine custom work and better quality paint jobs. In fact, magnificent show quality finishes were achieved with it by shooting on countless coats of paint and then sanding between each of them. Lacquer was

easy to apply, forgiving to work with, and could be polished to a beautiful, deep shine.

The problems with nitrocellulose lacquer were that it was rather brittle, and it needed constant maintenance to retain its luster and brilliance. Nitrocellulose lacquer was very low in solids at only 5–15 percent, which was the main reason so many coats were required for a good paint job. Nobody uses nitrocellulose lacquer for painting cars anymore, though you do run across it now and then.

Synthetic Enamel

The earliest cars were actually painted with brushes using oil-based enamel. Later, synthetic enamels were also the cheaper refinishing alternative in the old days after the manufacturers started spraying paint in the '20s because you merely had to shoot on two or three wet coats, and what you saw was what you got.

There were no catalyst hardeners back then, so synthetic enamel took a long time to dry. It would develop a hard surface fairly quickly, but it took about six weeks to cure. In fact, it never really quite dried completely, so you couldn't color sand and polish it. Also, synthetic enamel didn't dry in the true sense of the word, though it did off-gas the solvent reducers used to thin it. Instead of drying to create a durable finish though, it cured. The process involved a chemical reaction that took place rather slowly.

Synthetic enamel was also what they shot at those shops that would paint any car any color for $29.95 forty years ago. It was more flexible than lacquer, but many of the colors faded and chalked over time with exposure to the sun. Synthetic enamel isn't used much anymore because of the large quantity of solvent involved in shooting it and because of its long curing times. The solids (pigments, binders etc.) that remain after solvents evaporate off amount to around 10–20 percent.

Interestingly, the technologies involved in nitrocellulose lacquer and synthetic enamel were quite different. Lacquer literally did dry as we said before. It hardened and thickened as the solvents in it evaporated into the atmosphere, which took place almost immediately after the paint was sprayed. Lacquer didn't change chemically though, and if you rubbed it with lacquer thinner, even years later it would dissolve and come off.

Acrylic Lacquer

This was the finish preferred by General Motors from the 1960s into the '80s. Adding acrylic resins, (also called TPA, or Thermo Plastic Acrylic) to lacquer largely solved the problems of fading and

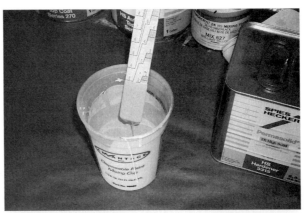

The new urethanes are pure chemistry and must be mixed exactly as specified for their molecules to cross-link and bond together properly.

The old lacquers could yield beautiful finishes but were fragile. Shrinkage and the vibration of the windshield on this classic have caused the paint to alligator.

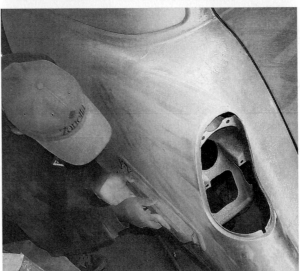

High-build primers are used to fill minor imperfections and ripples in panels.

chalking, though the finish it produces is still somewhat prone to chipping and is still somewhat brittle.

I have taken a bunch of trophies at prestigious shows with cars finished in acrylic lacquer during the last 20 years, but I sweat bullets all the way to the show and home again lest some truck flick a stone into a fender and shatter my labor-intensive finish like a pane of glass. Thankfully there are better solutions to the problem these days.

Acrylic lacquer is still available and is legal to use in many localities, but it is no longer the best finish for fine quality paintwork due to its relative fragility. However, if you are a novice restorer who is doing just one car, there are some advantages to shooting acrylic lacquer. If you don't mind the endless spraying and block sanding, a superb finish can be attained, and the paint is very forgiving to apply because you can adjust the ratio of solvent to solids without compromising the integrity of it.

Acrylic lacquer is also less toxic to breathe than the more modern two-pot paints, though it does pollute the atmosphere a lot more due to all the aromatic solvents it gives off. Just remember— before you shoot acrylic lacquer, make sure it is legal to do so where you live. The fines for using it where it is prohibited can be hefty. Solids for acrylic lacquer are somewhat better than the old nitrocellulose at 10–20 percent.

Acrylic Enamel

With enamel too, adding acrylic resins made a big difference as to durability and fade resistance. So much so that acrylic enamel became the standard for

the refinishing industry from the mid-'60s until the early '80s. Acrylic enamel is more durable and flexible than acrylic lacquer, but it still has a long cure time and it is not as durable as more modern paints. Solids are 15–25 percent, as sprayed. Acrylic enamel can be as toxic as the urethanes to shoot if you add the hardener that makes it tough and durable.

Acrylic Urethane

DuPont came out with the first polyurethane enamel in 1970 called Imron, and this type of paint has largely replaced earlier systems. Though it is still technically an enamel in that it cures chemically and by adding a hardener, the urethanes did away with the old problems of durability. In fact, the new two-pot, or two part paints are incredibly tough, and they can be color sanded and polished to look as good as the finest lacquer work.

These paints produce a brilliant wet look, and they are nearly impervious to solvents and chemicals. They are durable and flexible, and can be polished to award-winning perfection. But even urethane enamels have a few shortcomings. For instance, they don't work as well with metallics as

Polyester primers are essentially liquid plastic filler and make finishing a lot easier than it used to be.

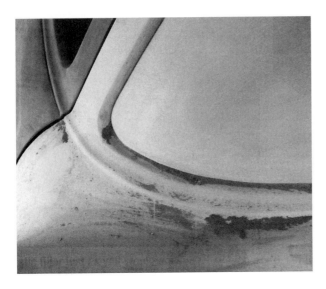

do base and clear systems, so they are generally used for solid colors, and they are rather expensive. On the other hand, many of the best paint shops shoot it almost exclusively.

The major drawback to urethanes is that the hardeners used in them contain isocyanates, which are cyanide. That's the stuff they used in the old San Quentin gas chamber, so you definitely don't want to breathe it or even let it touch your skin. A respirator mask with a charcoal filter is a must with any kind of paint, but with the new urethanes you need to take it one step further and wear a full plastic suit and use a positive pressure fresh air pack, or a system that allows you to breathe the air from you compressor outside the booth.

Base Coat/Clear Coat

There are a number of base coat/clear coat systems available including those that use a waterborne base coat, but they all use a urethane clear coat. Clear coats began to become popular around 1981 as a way of enhancing the luster and depth of metallic paint jobs. Shortly thereafter, the auto manufacturers started using it on all their finishes.

Today's base and clear coats vary in what they use for a base coat, with most of the OEM base coats being waterborne, but they all still use a clear solvent-borne urethane top coat at this writing. Base coat/clear coat systems are easy to shoot, though you must follow the base coat with the clear coat at the right interval in order for them to laminate properly, unless you use a waterborne base coat which will bond mechanically. Be sure to follow the paint manufacturer's instructions to the letter when shooting any kind of paint, and especially base coat/clear coat finishes.

Waterborne Paints

Several years ago waterborne paints were pushed heavily because of their low VOC (Volatile Organic Compounds). These paints are not actually soluble in water as one might imagine, so they are not water-based, but waterborne. Purified water is just used as a vehicle to get the paint onto the car. They are no more difficult to spray than solvent-based paints, but they do need an epoxy primer under them to protect the metal, and they are not as easy to repaint as some other systems. The paint itself is somewhat dull, so a urethane gloss topcoat is needed to make waterborne paint look its best, but more on that in Chapter 19.

Primers

There are a lot more choices these days as to what you want under the paint to make it stick, and to even it out and help protect the metal too. The first category is the metal etching primers designed to develop a very tight bond to the metal below. After washing bare metal parts with a degreaser to get any residue of grease or oil off, and going over the whole part carefully with a wire wheel to remove any rust from the pores of the metal, you are ready for primer.

Metal-Etching Primers

Metal-etching primers use acid to etch the primer into the metal below, but they are not high-build primers. They have great adhesion, almost like the E-coats on replacement parts though, and help protect parts until you are ready to shoot on a high-build primer, block sand and shoot on the color coats. Some of the toxic ingredients such as lead and chromate that made these products so effective have been removed in recent years though, so they aren't quite as good as they used to be.

Epoxy Primers

Two-stage epoxy primers applied over clean bare metal are very effective for moisture-proofing and protecting them until you can shoot on high-build primers and color coats. Epoxy primers etch into the metal too, and bond very well. If you see an epoxy primer that is single-stage, avoid it because it is not a true epoxy primer. You can use epoxy primers over etching primers as well, to keep the etching primer's acids from causing problems with the color coats.

Polyester Primers

This is the greatest thing since electronic fuel injection because it makes final finishing so much easier than it used to be. High-build polyester primer is essentially liquid plastic filler and allows you to sand out any minor unevenness in panels as well as bodywork. It still won't save a bad panel, but final finishing of plastic filler is less critical with polyester primer.

Acrylic Lacquer Primers

For those still shooting lacquer, you must use only lacquer primer. Any manufacturer of acrylic lacquer sells an array of primers that are compatible with their paints. Metal-etching primers can be used, as can high-build primers, but make sure you get the right stuff for the job. It is important with any paint system to stay with one manufacturer for the entire paint system you apply, and it is especially critical that you stay with one type of paint, such as lacquer, enamel or urethane.

Urethane Primers

These are used as the final coat before applying color. Whether you use a base coat/clear coat or a single-stage urethane paint, you will want a coat of urethane primer underneath, and it too should have a final block sanding using 500-grit paper (800-grit before waterborne color coats). You will want to sand the final primer to even out the finish, but more importantly, you will want to avoid a final finish that is more than 7 mil thick. That way you will avoid cracking due to flex and excessive shrinking of the finish due to temperature change.

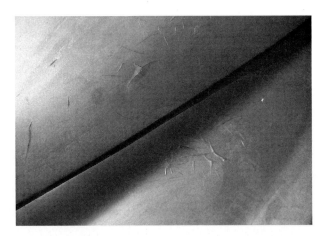

This is a typical case of one type of paint not adhering to another. Strip old paint off if you don't know what it is, and only use products from one paint system exclusively while painting a car.

Chapter 19
Waterborne Paint Tech

Henry Ford would have welcomed waterborne paint because it takes less of it to color-coat a car.

To many of us, the new kid on the block is waterborne (not water-based which would also be water soluble) automotive paint, although it is not really that new. The auto industry has been using it for years; but increasingly environmentally conscious local and state authorities are mandating it for repair, repaint and restoration applications too. In fact in many localities, including where I live, the only kind of color coat you can legally shoot is waterborne. However, the primers and final clear coats that are used with the waterborne color coats are still the old-style solvent-based paints at this writing, though that will most likely change too.

So how do they stack up? Very well, as it turns out. They are durable, brilliant, and fairly easy to apply once you get the hang of it. David Escalante of Custom Auto Service in Santa Ana, California, restorers who have produced Pebble Beach winning cars, says much of the difficulty painters have with waterborne paints is attitude. Painters get good at one system and they want to stick with it. Waterborne paints spray differently, lay differently, and dry differently, but learning to spray them is not difficult. Escalante says any experienced painter can shoot them, but it is a good idea to take a class if one is available.

Why Waterborne?

The reason for the change from solvent-based color coats is to reduce the total V.O.C.s, or volatile organic compounds, being put into the atmosphere. Now you might ask, if the primers and clear coats are still solvent based, what is the point? It is simply this: By going with waterborne color coats, the overall amount of pollution generated when painting cars is significantly reduced. But keep in mind that waterborne paints are not harmless. You still will need to wear a paint suit, gloves and a proper fresh air respirator when shooting them because they still contain toxins such as acetone.

The Differences—Because of their slow drying time, fans are a virtual necessity to move air over the painted item, and your spray booth must be squeaky clean. Ditto the air going into the booth, which should also be as dry as possible. And your paint gun needs to be made of stainless steel because water corrodes ordinary steel. Also, you must use a good filtering system for your air supply, because waterborne paints are more susceptible to fisheye from contaminants.

It is critical to remember that waterborne automotive paints must be stored, shipped and sprayed in a temperature-controlled environment too. And you can't put them in a paint shaker at the store to stir them up because doing so will cause them to foam up permanently, and you can't paint

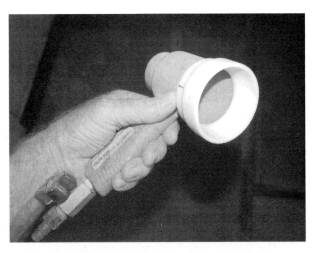

Handheld fan is fine for small jobs, and Eastwood sells portable stands and racks for a modest price.

Kye Yeung at European Motor Car Works shows us the fans and lights in his state-of-the-art spray booth.

Yeung also points out that the air going into your paint gun must be absolutely clean and oil free, hence the three heavy-duty filters.

with foam. Instead, just stir them gently and thoroughly before putting them in your paint cup. You won't be able to store any excess paint for future touch-up either. But that is not usually a major problem these days, thanks to computerized paint matching systems.

Advantages—Obviously the fact that they are far less toxic and harmful to the environment is a big plus. And the slow drying time does allow you to spot them in a lot more easily. Waterborne paint is also great when doing a scuff and repaint too because it is compatible with lacquer, enamel and urethanes. That's due to the fact that it links

mechanically with the paint under it, rather than chemically. It adheres beautifully to just about any surface. And it is a real plus that there is no problem with having to shoot multiple coats or urethane topcoats within a certain time limit, as with solvent-based chemically bonding paints.

In other words, you can shoot a coat one day, and another coat two days later without problems. On the other hand, if you don't clear-coat solvent-based color coats within twenty-four hours the clear won't stick, and will peel later. And most importantly, waterborne paints have proven themselves to be as durable as any other paint system provided they are applied correctly. And best of all is, you need less of it to paint a car. Dealers say a mere two quarts of the stuff will do an average car, and with the cost of auto paint these days, that is a blessing.

Applying Waterborne

To begin with, all bare metal must be carefully sealed, including seams, grooves, lap welds and crevices. If they are not sealed, water will run down into them and cause rust. And once rust begins, it will grow like cancer until it is removed. Consult your automotive paint supplier as to which sealers and primers to use. And though waterborne paint suffers fewer compatibility problems, it still pays to stick with the same paint system such as Sherwin Williams or PPG all the way through the process.

High-build, or other primer can then be sprayed, and the car is then block sanded to eliminate low and high spots and other irregularities just as you would do in preparation for solvent-borne color-coats, except that you will want to take your final block sanding a little further according to David Escalante. He says that he goes all the way to 800 to 1,000 grit before applying waterborne color coats. The car is then carefully wiped down with Prep-Sol or another degreaser before going to color. And through the entire process Escalante is extremely

Paint guns for waterborne paint must be made of stainless steel, aluminum or plastic to avoid corrosion.

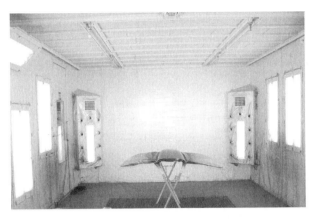

State-of-the-art spray booths used for waterborne paints have fans, as well as plenty of light, and are kept at a uniform 78 degrees.

European Motor Car Works does a lot of high-end exotics and uses waterborne color coats with confidence.

careful to avoid oil or grease.

To shoot the color-coats, open the spray orifice to full, in order to get about an eight inch spread, and set your pressure at about six pounds (or a little higher if you are a fast painter). The paint will look rather terrible at first because waterborne goes on differently, but as it dries it levels out nicely. Shoot the paint from about twelve inches away instead of the usual hand-width distance, because if you get too close you will develop an effect similar to fisheyes.

As for drying, at European Motor Car Works in Santa Ana California, Kye Yeung, who paints Jaguars, Lotus's and custom show rods with legendary results, tells us that they keep their spray booth heated to 78 degrees Fahrenheit at all times, and proper fans are installed at the four corners of the booth to move lots of air so the water will evaporate quickly. But both Yeung and Escalante ply their trade in Southern California, which is generally warm and dry most of the year. In the east and in colder wetter climates, humidity would be more of a concern, so heaters and dehumidifiers would be in order.

What You Need—The day when the home hobbyist could paint his car in his garage just by

wetting down the floor and donning a respirator mask are gone forever. Fist of all, modern paints are extremely sensitive to contaminants, humidity and temperature, and secondly, unless you live out in the country, you are likely to run afoul of authorities concerned about the atmosphere and environment. To do the job properly you will need to rent a spray booth, and that may not be as difficult as it might sound.

If you are lucky enough to have one nearby, you can take a class at the local junior college and use their booth. Or you can get to know the proprietor of a local body shop who would be willing to let you use his spray booth on the weekend. And finally, in some localities, there are spray booths available to rent commercially.

When shooting waterborne color coats you will want to spray you car on a warm day in a clean, dry booth because contaminants are a much bigger problem for the slow drying waterborne paint. However, you could even use big household fans to dry your work in a pinch. (Never use such fans for solvent-borne paint because you risk explosion and fire if you do.) The Eastwood Company also sells fans and racks on which to mount them, at modest cost, and small, handheld fans are also available for drying smaller components.

As we said before, ordinary steel spray guns will not suffice for spraying waterborne paints due to corrosion. Instead you will want a paint gun such as the DeVilbiss FLG 647 WB with the 1.3 mm spray tip, which is also available from Eastwood. All of its components are made of aluminum or stainless steel.

You will most likely also want to pick up an adapter for the gun that allows you to use disposable paint cups. One of the chores I always hate when cleaning a paint gun is sopping out the cup. And even though you can clean your gun and cup quite nicely with ordinary distilled water, there is also a gun cleaning wash available from automotive paint stores. It is extremely important to clean your gun meticulously immediately after using it because of that mechanical bond thing with the paint.

The disposable cups are also nice because they have graduated measuring tables printed on them that allow you to mix paint and reducer etc. precisely. There are full size cups available for big jobs, and smaller ones for touch up too.

Once you have shot on the color coat, if you need to sand out any problems, use only open coat, dry sandpaper so you won't run the risk of corrosion. Five hundred grit paper will generally suffice, and should not create sand scratches in the clear coat when it is applied.

You will want to do your masking with 3/4" or 1 1/2" wide 3M green masking tape developed for waterborne applications, as well as Scotchblok masking paper, which is gold in color. And if you are two-toning or doing custom work, pick up some of the purple fine line tape as well.

In conclusion, practice makes perfect, and attitude is everything, so spend a little time practicing on old parts to get the hang of it, before trying your luck with you precious ride. Getting used to waterborne is no more difficult that it was for painters of ten or fifteen years ago to get used to HVLP (High Volume, Low Pressure) and you will be doing the world a favor by switching.

Robert Escalante at Custom Auto Service shows us the scrupulous records he must keep of the paint used in his restoration shop.

Disposable paint cups are great because you don't have to clean them, and they have measurements marked on them.

Chapter 20
Block Sanding

We made a couple of little fixes after the first coat of high-build primer was applied, then applied a misted guide coat to show us where more fixes need to be made.

Block Sanding

Block sanding is the secret to obtaining a perfectly flat surface for the final color coats so highlights will play flawlessly along each panel. The task isn't tricky, but it takes a fair amount of time and care. Also, there are a couple of ways you can go, and it is mainly a matter of personal preference as to which you choose. I'll explain one very good method in common use first, then I'll go into Tom Horvath's favorite technique. This first approach is ideal if you are going to be painting a car over a period of weeks and will need to store parts unpainted for some length of time.

Diagnostic Disassembly

You may ask, "Why take your car apart to paint it in first place?" And to that I would say because it is the only way to completely rid it of rust, dirt, and previous damage, and it is the only way to adequately protect such surfaces as the inner skins of fenders and doors. Whether or not you decide to go so far as to take the body off the frame and do a total restoration should depend on how much time and energy you want to devote to the project, whether the car really needs such an all-out effort, and whether you want to completely rip out the existing interior—assuming it is still good.

In my mind there are very few cars that need to have their bodies lifted from their frames, but most older cars that need a total, first-class professional paint job should be disassembled. The front clip consisting of fenders, hood, grille, inner fenders and mud pans should come apart and be painted. If the doors need a good deal of work, they should come off too, as should the trunk lid if it is damaged.

As I said earlier, cars are easy to put together, but that is only true if you have everything at hand, properly labeled and cleaned. If you have to spend hours looking for the right fasteners and wondering where that little piece of trim went, the job can take forever. So when you take a car apart, do it in an orderly fashion. Take photos. I use a small digital camera and a laptop computer for storing my photos. Take notes and label all fasteners. Put them in plastic Ziploc bags and label the bags.

When removing doors, hoods and deck lids, scribe around their hinges so you can relocate them correctly, and save any shims you find under them. If you don't do these things, your panels will be difficult to position and will not lie flat. The more care you put into taking the car apart the easier it will be to put it together when the time comes.

Method One

Whether or not you are going to paint your car whole or in pieces, the following methods for block sanding apply.

Block sanding takes off the high spots first, leaving the low spots which are obvious places to be filled or fixed.

This is what the trunk lid looks like after careful block sanding with 400-grit wet-and-dry sandpaper and plenty of water.

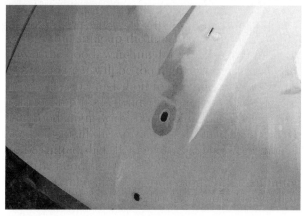

Low spot in hood where ornament was attached probably occurred as a result of people closing it too aggressively. This can't be filled with plastic because it would just crack out. A little tapping from the back will fix it.

Shoot the Primer—Spray on two or three coats of the epoxy primer chosen from the paint system you will be using. When that has cured (I'd give it eight hours indoors in a dry environment), shoot on about three coats of high-build primer to help even out the surfaces of the body panels. Contrary to what you might believe, a thick paint job isn't more durable than a thin one. In fact a thick finish is more prone to cracking because of flexing and uneven heating of the surface in the sun. What you are trying to achieve with the primer is the most perfect possible surface for the color coats, and that entails sanding most of the primer away.

Only use the high-build primer the manufacturer of your paint system recommends, and after shooting on the specified number of coats, let the primer cure for a few days indoors in a dry climate so the paint can off-gas properly. Trapped solvents and aromatics can cause paint to crack and delaminate, so you want to give them time to evaporate out. You will know when the primer is completely dry because you will not be able to push the edge of your thumbnail into it and make a dent.

Give It a Guide Coat—A guide coat is a light fog shot onto the primer so that when you block sand it off you can tell where any high or low spots or blemishes are in the bodywork and correct the problems before going further. Cheap aerosol paint is good enough for a guide coat because you will be sanding it all off anyway. Just mist the paint on lightly and let it dry completely so it won't clog your sandpaper.

Now place several sheets of 400-grit wet-and-dry sandpaper in a bucket and put in a couple of drops of liquid dishwashing detergent. Fill the bucket with clean water and let the sandpaper soak for about 20 minutes so as to soften its edges. Fit the sandpaper into a rubber sanding block, then

thoroughly wet the surface to be sanded before going to work. Keep the surface you are sanding wet all the time. A running garden hose is good for this, or use a big sponge to keep the water coming.

Work back and forth in an X pattern to avoid making low spots or grooves in the primer. Don't put any muscle into it. Just let the sandpaper do the work. Keep dipping the sandpaper to prevent it from clogging, and change it when it stops cutting. Once the grit gets dull and you have to start bearing down to make it work, from then on you will only make grooves in the finish.

Thanks to the guide coat, high spots will show up first. When you find them, do a bit of work with a picking hammer or file to take care of it. And of course, if you find a bigger problem, you may need to use a shrinking hammer or dolly to draw it in. A pro will have dealt with all but small irregularities before the primer coat, but a novice may not notice such problems until block sanding. If you find problems that require more filler, sand off all the primer to bare metal in that area and make the repair. Touch it up with epoxy primer, then shoot on primer-surfacer again.

As you sand, an occasional low spot will show up. If you can tap these out easily, do it, but if the low spots are shallow enough to be repaired by a thin layer of filler, that is probably the best approach.

There are a few fixes around the trunk hinge that need a little work. More panel beating and a little filler, followed by another coat of high-build primer will fix that.

After spraying the body of this sedan delivery with polyester primer, Tom shoots on a guide coat with aerosol paint.

When the guide coat dries Tom uses a sanding block to sand in a crisscross fashion to even out the surface.

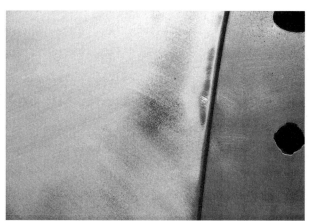

No matter which method you use, you will break through to bare metal in places. These must be primed and sanded again before painting.

Again, you must sand off all the primer to bare metal in the area to which you apply the plastic filler. Once you have the repair made, spot prime and block sand again.

Fix small irregularities and pits with glazing putty. Don't try to fix actual dings or dents with it though because it is not intended for such use. Putty can be applied over primer. Just let it dry thoroughly, then sand it flat before going on.

My little Morris you see in the photos is a nice rounded car, so block sanding it is not difficult. Rounded surfaces make block sanding bodywork easier because they don't develop or reveal ripples the way large, flat surfaces do. Very tight curves can be a challenge though. Find an old radiator or heater hose and wrap your sandpaper around it to make a flexible sanding block.

Stay away from sharp edges completely while sanding. Paint tends to be thickest in recessed grooves and thinnest along the edges of doors, hoods and trunk lids. In fact it is a good idea to tape these surfaces off with masking tape to avoid breaking through to bare metal. You will definitely want to shoot the car with more high-build primer and block sand it again at least one more time, so breaking through on the first primer coat isn't that critical, but you can't have any metal showing when you shoot on the color coats.

With things like color sanding you have to try to reach an almost Zen meditative state of mind while you work. Close your eyes periodically and run the palm of your hand over the surfaces to feel for any low areas or imperfections. Surprisingly, you can feel even slight imperfections if you pay attention. Your spouse might look askance at you for this, but it can save you problems in the color coat stage. There is no point to rushing the job. Any imperfections you find at this stage will be much easier to deal with now than after you shoot on the color coats.

Take your time and examine each part of the car with the same care. Any unnoticed dings, surface irregularity or ripples will only become much more evident after you shoot on the color coats. It is at this block sanding stage that great paintwork is accomplished. Only after you have sanded most of the primer away and gotten the surface absolutely smooth are you ready for the next step.

Tom's Technique

Tom Horvath is a pro with a big shop and an experienced staff. He doesn't let panels sit around for weeks unpainted, even though he demands

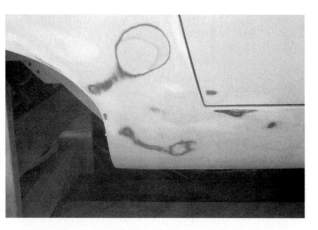

Sanding over coats of primer, filler and old fixes can be time-consuming, depending on how well you prepared the surface in the first place.

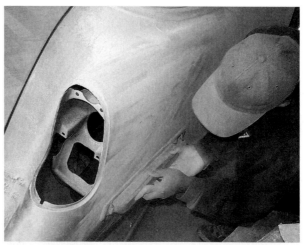

Long, broad strokes with a long sanding board are the way to even out low crown and flat surfaces.

Clemente has a collection of small blocks, dowels and pieces of hose for sanding different surfaces.

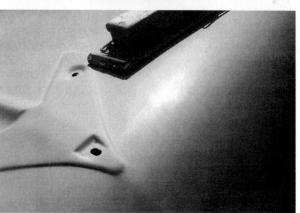

When you are finished block sanding with 500-grit, the primer almost shines.

exquisite Pebble Beach perfection in all of his paintwork. Of course, Tom also uses guide coats as outlined above, so we won't go over that again. Here's his approach:

Tom takes every panel down to rust-free bright clean metal before painting. As soon as the bodywork is completed, Tom then shoots on polyester primer. That's pretty much like shooting on a very thin coat of plastic filler. It's essentially the same stuff. The only problem with polyester primer is that it is not moisture-proof. And because it is not moisture-proof, you must shoot a sealer on it before putting any kind of top coat over it—especially waterborne.

Once the primer has cured (about 8 hours) he block sands using air-powered sanding boards and 220-grit dry sandpaper. He then goes over the car carefully and makes any little fixes required. This is a lot easier using polyester primer because plastic filler sticks to it just fine. You don't have to strip to bare metal an area that needs work again each time.

Next, he shoots on more primer and block sands again, only this time he goes to 320-grit paper. Then Tom shoots on a third heavy coat, goes over it with the 320-grit again, and finishes with 500-grit to produce a surface that is so perfect it almost has a dull sheen even without a color coat (for waterborne color coats you will need to go all the way to 800-grit paper before painting). Of course, each step of the way Tom and his crew are checking carefully for any slight irregularities and fixing them.

The primer system you use isn't as important as the care and attention to detail you lavish on the job. As we said before, use the palm of your hand with your eyes closed to feel for problems. Check each panel in good light either indoors or out (just inside a big garage door on a bright day is the best)

and look at it from every angle to help spot imperfections. Only go to color coats when you have done your most critical inspection and determined that you have the surface you want, and then shot on a sealer. That's because any imperfections will only look worse—not better—after you shoot on color.

Chapter 21
Masking & Final Prep

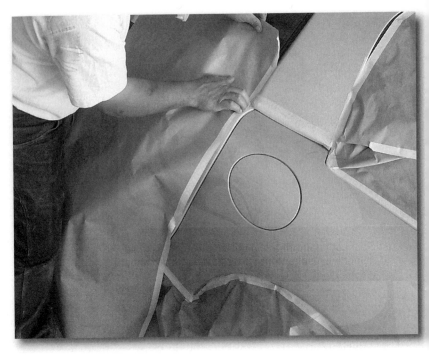

Masking is not difficult, but there are some tricks of the trade that take a little practice to perfect.

Before you even consider unlimbering your spray gun, make sure everything is clean—including you. Studies have shown that much of the contamination in spray-booth paint work comes from the painter. So that means you can't spend the morning grinding out plastic filler and sanding high-build primer, then lightly brush yourself off, step into the booth and start painting. Not if you want a decent paint job.

Scientific studies of the sources of paint defects at the factories showed that 60% of the problems were caused by dust particles, lint, hair, dandruff, powder, cotton fibers and other contaminants carried by workers into the spray booth. Anyone entering a paint booth should wear clean shoes, lint-free coveralls and a hair bonnet. And for safety's sake they must also have a source of fresh air, either from a system such as Survivair makes, with its own air pump, or a system that cleans and filters air from the shop's compressor.

You could get away with more in the old days with nitrocellulose or acrylic lacquer, but the newer paints don't dry virtually on contact the way lacquer did. The flip side though, is you don't have to shoot on 10 coats of the new urethanes and sand nine of them off to get a good finish the way you did with lacquer. Old-style acrylic lacquer is still available and legal in some localities, and it still is more forgiving to use than modern paints, but it has several disadvantages, among them being that its solvents disperse into the atmosphere and cause pollution, and the fact that lacquer is rather fragile in service.

So unless it is legal to shoot lacquer in your area, and unless you are willing to content yourself with obsolete technology, you, your car, and its surroundings will need to be surgery-room clean to get the best results. Besides, even lacquer requires a reasonably clean and dust-free environment. Here is a recipe that will give you good results no matter what you shoot:

Spray Booth

Is it possible to shoot automotive paint without a booth? Yes. But chances are it isn't legal in your locality, and unless you erect something very similar to a spray booth in which to work, your finished job will be marginal at best. You can contain the overspray and limit contamination by hanging plastic tarps from a temporary structure made of wood or pipes, and you can hook up a fresh-air pack outside the structure, but you still have to deal with the fumes.

In the end, if you are just getting started or are an amateur, it will probably be better to find a local paint shop that will let you rent their booth in the evening or on the weekend to shoot your car, than it would be to erect a temporary booth. Other possibilities are to enroll in an automotive collision repair class at the local city college, or if you live in a larger metropolitan area, look in the phone book for spray booth rentals. A booming business is developing in booth rentals, so you might just get lucky.

Painting your car in your garage is definitely not a good idea either. Again, it is very likely illegal in your locality, and there is the possibility of explosion and fire. Electric appliances and water heaters can kick in anytime and ignite fumes. And unless you leave your garage door open and install big fans, there won't be nearly enough ventilation anyway. Nor will there be anything like the amount of light you will need to do a good job. And unless your garage is finished off, there will be all kinds of dirt, dust and spiderwebs up in the rafters and nooks and crannies everywhere. Garages were definitely not designed to be spray booths. For more on spray booths, see Chapter 17.

Prep the Booth—Make sure the ventilation system is working, its filters aren't clogged, and that you have a ready source of compressed air. Turn on the compressor and let it pump up. Sweep out the booth, but try not to whip the dust and dirt into the air. Never blow out the booth with compressed air. All that accomplishes is to put filth and contamination in suspension and swirl it around.

When the booth has been swept thoroughly, mop it down using a bucket of soapy water, rinsing your mop frequently. Some painters wet down the floor of the booth, but this has its downside in that the moisture can be splashed up on the car while you are painting, and any electrical apparatus nearby could cause a shock.

Prepare the Car

Whether you paint your car in pieces or all together, be sure to clean and prep it or its components outside the booth. All sanding and repair work must be completed before you begin prepping. If you do come across something that needs a little more block sanding, do the job outdoors away from your work area. No sanding should ever be done inside a spray booth. This may seem obvious, but plenty of people do it when they are in a hurry and in doing so defeat one of the main purposes of the spray booth, which is to maintain cleanliness.

Before you take them into the paint booth, take the car, or its parts outdoors and blow them off completely with compressed air. Be sure to get down into grooves and pinch moldings as well as down in holes for door hinges and handles. You can bet that any dirt or dust lurking in these areas that you don't get out now will be dislodged by the air from your spray gun and stick in your new finish. Take your time and be thorough. A little extra

A taping machine makes masking a lot easier. If you don't have one, you'll need to spread paper on a workbench and apply the tape carefully along its edge.

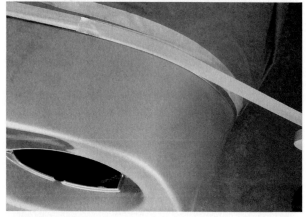

Stick tape down and gently stretch it as you pull it around. Don't mince along in short sections, because you will cause crimping and unevenness if you do.

This is double-taping. First the edge to be masked is taped, then paper with tape on it is applied to that.

effort at this point will save you from heartbreak and frustration later.

Next, get plenty of lint-free rags, make a pad about eight inches square and about six layers thick and pour a little DuPont Pre-Kleeno into it and wipe the car down carefully. This is mainly to get rid of any grease or oil that might be present— especially if you have been touching the block sanded surfaces with your bare fingers. Keep changing the pad around so you are using clean

Folding tape like this is a good way to make up tape for grooves and for back-taping.

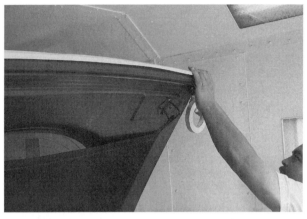

Folded tape is applied along a groove to avoid a hard line at the edge of the parts, but seal against the gasket underneath.

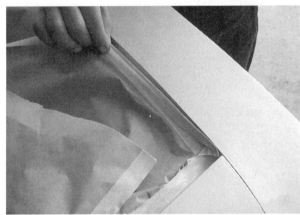

Clemente at Tom's Custom Auto Body double-tapes around this Corvette rear hatch in order to get a good clean line.

cloth and fresh cleaner at all times. Don't be stingy with the solvent either. There is no point to just pushing the dirt and grease around.

Mask the Car—Masking is the part of prep that separates the pros from the novices. Nothing says "cheap paint job" like overspray and paint on rubber moldings, trim and upholstery. Masking isn't difficult, but there are a few tricks to it, and the right equipment helps too.

By the right equipment I mean a taping machine. Nothing fancy. These simple and inexpensive devices put the tape on the edge of the masking paper uniformly and they leave just the right amount of tape exposed. You can mask without one of these but it can be a little tricky. It takes practice to avoid tying yourself in knots with sticky tape and then going into a whirling dervish routine to keep the tape from bonding itself to the paper in places you'd rather it didn't.

If you don't have a taping machine, you can spread the paper on a workbench and then add the tape. This takes a little practice, but it works just as well. Keep the tape stretched straight, stick it down at one end then the other, and then press it into place on the paper. Don't try this working on the floor though, because you will pick up dust and dirt no matter how carefully you prepped the booth.

Be sure to use only professional quality masking

paper available from your local automotive supplier. Newspaper is full of lint, dust and paper particles and has those ragged edges that shed debris, so it has no place in the pro's shop except to wrap fragile parts to keep them from breaking.

Have plenty of masking tape and paper on hand before you start. Paper and tape are cheap but getting overspray off is a big pain. Anything you don't want painted must be masked. Don't assume you can just be careful to avoid things you don't want painted. Overspray floats around in a very sticky state and will settle on everything. And don't even consider smearing grease or Vaseline on items to keep paint from sticking because sure as can be you'll also get it on things you DO want the paint to stick to and big fisheyes will be the result.

Tape Tricks—There are pros in production paint shops that can mask a car in twenty minutes and do a pretty good job, and in a production shop, time is money. But if you really want to mask a car correctly—especially if you are a novice, you'll need to double-tape it. By that I mean you will run tape around, for example, the back window rubber, then attach your taped paper to that, rather than trying to do the job in one swoop with just the taped paper.

Another important trick is back-taping. That's where you pull the paper back over the tape and fold it down. This is a good technique to use when you need to feather an edge. Another trick is to fold about 1/3 of the strip of tape over on itself to make a lip. You can then use this tape along, for instance, trunk or hood seams where you don't want the tape to actually touch the parts. You can also pull back this folded tape and double-tape against it with paper. Here are some taping basics:

Pull as much off of the roll as you are going to need to cover a span, but don't tear it loose from

Just to make sure no overspray gets into the interior of the car, Clemente tapes in paper and folds it back before closing hatch.

Be sure to tape down all of the paper seams so air and overspray can't get under it during painting.

Double-taping behind fender bead is the way to go if you will not be painting up into wheel wells.

Take the wheels off and wrap the brake disks and suspension if you need to get inside wheel wells.

Tape off headlight openings from behind so you can paint edges of opening.

the roll. Now stick the tape down at one end. Use the tape roll as a handle and guide the tape into place with your outstretched arm. Don't try to put down tape a few inches at a time and mince it into place gingerly. You'll end up with sloppy edges and crimps that will let paint in.

Masking tape was designed to go around gentle compound curves. If you stick it down, then pull out a couple of feet or more and keep it stretched tight you can actually do compound curves with it. In any case, stretch the tape as you work and pull out most of the little wrinkles that allow it to go around curves so paint won't get under the little raised areas.

If you don't get a nice, straight line and adequately cover what you are trying to mask, pull up the tape, throw it away and try again. If you stick old tape down again, it will not hold as well and can give you problems later when you least expect them. Check each edge carefully before going to the next one.

When you get to a sharp corner, overlap the tape. Don't expect ragged corners to keep paint off of parts. Any loose tape that is sticking up can deflect paint as you are spraying and cause runs and holidays, so stick them down. Finally, be sure to tape off all the seams in the paper completely, lengthwise along the seam. Don't expect a small piece of tape to hold the paper in place while you are shooting paint. Overspray will blow under loose paper and get into areas where you didn't want it.

Wheels

Wheels are another challenge. There are a couple of good ways to deal with them. One is to put the car on jack stands, take the wheels off and set them aside. This is a good idea if you will need to shoot paint up in wheel wells. Just make sure you mask off the chassis, drums or disks, and any other components you don't want painted. The other way to handle wheels is to double-tape from behind the fender bead, then pull the paper down over the wheel. Be sure to wrap the tire completely all the way around its circumference though.

If you have removed headlights, taillights and the like, you will need to mask behind them because overspray can get into such areas and get all over the tires, wheels and wiring. Paper can be taped in behind headlight openings, and it should be stuffed into holes for taillights, making sure the paper doesn't touch the edges you want to paint. Don't forget to mask off those chrome-tipped exhaust pipes either.

When you have the car wrapped as well as you can, take one last, slow walk around it and examine it for details you missed. If everything looks right, you are ready to put on your painting suit, go over the car lightly with a tack rag, and start spraying.

Stuff paper in behind taillight openings and tape it into place to keep over-spray off of wheels.

Don't forget those exhaust pipes either. Washing overspray off of metal parts isn't as easy as it was in the lacquer era.

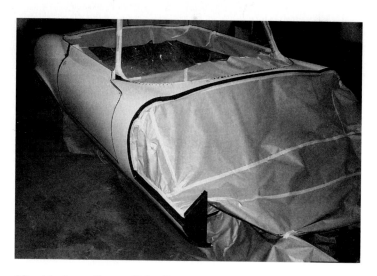

Bruce sealed the floor and interior panels first, then shot door jambs, cowl, and fender wells before final block sanding.

Chapter 22

Shooting Paint

Because I am of a certain age, I can remember when painting a car was simply a matter of scuffing the old surface, masking off the brightwork, wetting down the garage floor and shooting on a few coats of lacquer. Nobody even bothered with a respirator, but you usually stubbed out the unfiltered cigarette you were smoking before starting to spray. But auto painting has become a lot more dangerous these days.

To those daring types out there who might still actually consider shooting paint this way, don't. Some of the painters I'm talking about didn't live long enough to tell their stories, and some of those who are still around aren't in great shape. Besides—as we said before, and will keep saying until everybody gets the message—paint has become a lot more toxic in recent years, meaning it can kill you in a hurry rather than just causing fatal damage over a long period of abuse.

Most things to do with painting a car have changed completely in the last 20 years, although a few have not. You always did have to be thorough, methodical, and patient. And you always did need to start with a clean, rust-free surface and spray in a clean dust-free area. But the paints a few years ago were a lot more forgiving than the new stuff, and especially waterborne color coats. You always did have to be careful of dust and contaminants, but these days you have to be downright obsessive, especially if a nearly perfect finish is your goal.

After block sanding and blending around painted areas, the Jag was masked off for final finish.

My buddy Bruce showed me how he shoots a car when a show-quality finish is the goal. He has been busy converting a Jag E-model coupe into a roadster and has finished the body to his satisfaction. Countless hours of aligning, filling, priming and block sanding have finally resulted in a body befitting a Jaguar. Now all it needs is a shimmering, deep finish in the proper British Racing Green selected by the customer.

Bruce Haye did a bit of touch-up with primer then sanded before color spraying.

Before any paint work was done, Haye went over the whole car with Prepsol to remove any oils or silicone.

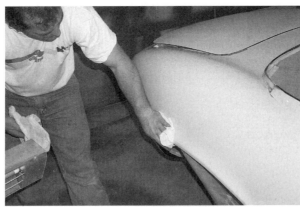

The final step before shooting color is to tack the entire body and any masking to catch any dust that might have settled.

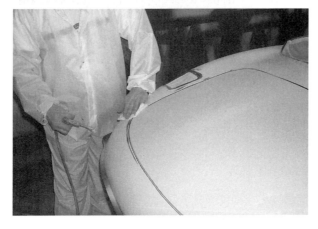

A full plastic suit and gloves is required to shoot urethane because of its toxicity.

Fresh air from outside the booth is piped in and filtered. An ordinary respirator mask is not enough these days.

Jambing

Before Bruce block-sanded the outer body, he undersealed the floor pans and driveshaft tunnel inside and out. Then he masked the car carefully and shot the cowl, the interior panels including under the dash, and the doorjambs and trunk with a couple of coats of color. That way there would be no overspray, rough spots or runs as a result of trying to paint these areas when the rest of the body was sprayed. Then the body was wheeled aside to cure for a few days.

Next, Bruce shot on a guide coat and block sanded the body one last time, blending the areas where he had shot on color. He used a rubber sanding block and worked in a crisscross fashion so as not to make grooves, then finished with a light hand-sanding with 400 dry, open-coat sandpaper. Then the entire body was taken outdoors, blown off and wiped down.

Touch Up

During the final sanding a couple of small high spots appeared, so Bruce touched them up with primer, then blended them carefully with fine sandpaper. After that, the body was ready to spray. Bruce had thoroughly cleaned and pre-heated the spray booth to 70 degrees before the car was wheeled in so the paint would have a warm, dry, dust-free environment in which to cure.

Final Wipe Down

Bruce dampened the floor to prevent any unseen dust from blowing up into the paint. This is an old

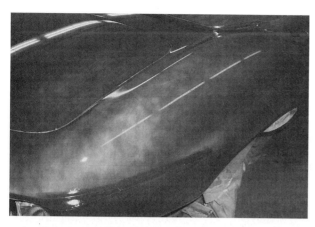

Bruce shot on a tack coat, giving every part of the car the same careful attention.

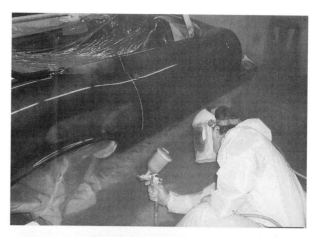

When shooting the final color, Bruce did the hard-to-get areas first, then shot the rest of the body.

The old cat already looks magnificent in British racing green, but will be rubbed out to dazzling perfection later.

trick that fell out of favor a few years ago, but is fashionable again, especially since contamination can't be as easily sanded out as it used to be with the old-style paints. This step is a matter of choice. Not everyone does it, nor is it necessary if your booth is really clean. If you do decide to wet the floor, just be very careful to unplug any power cords, and make sure no stray hot wires are anywhere near it.

Next Bruce wiped the whole body down one final time with a soft, clean cotton cloth and Prepsol to remove any last bit of oily contamination. He went over every inch of the body with the same careful attention to detail. Not even the hard-to-get-at, tucked-under panels were slighted. Bruce was generous with the solvent, and kept turning and changing his rag to make sure he wasn't smearing contaminants around.

When the entire body had been wiped clean and had dried completely, Bruce hooked up a nozzle to his compressed air gun and got out a tack rag and neatly folded it in squares. Tack rags are sticky because they are impregnated with a type of non-hardening varnish that is necessary to remove any last particles of dust. Then Bruce used a light blast of compressed air to dislodge any stray particles from cracks and crevices. He then went over the whole body of the car again carefully and gently, so as not to smear any of the sticky varnish in the tack rag onto the finish.

Mixing Paint

As we have stated elsewhere, the paint used for quality finishes nowadays must be mixed in precise proportions. These new urethanes are pure chemistry. Adding solvent to them doesn't just thin them, it also breaks down the molecular cross-linking that makes the paint durable. Bruce used a

special aluminum, graduated mixing stick and a plastic graduated pitcher to get the formula right. Then he carefully and thoroughly stirred his paint, and strained it into the cup of his spray gun.

Safety Issues

We talk more about this in Chapter 1, but briefly, you must keep in mind that modern urethanes and even acrylic enamels have isocyanates in them for hardeners, which are a close cousin to cyanide. That makes them deadly to spray without full body protection (it can be absorbed through the skin) and a supply of fresh air from outside the spray booth. The solvents in the old-style lacquers weren't good for you, but this new stuff can get under your skin and cause kidney failure and even death in a short period of time.

In the ceiling of Bruce's spray booth are massive filters through which outside air is passed from vents down low to produce a positive atmosphere inside the booth so dust cannot drift in. Bruce checked his gear carefully going down a mental checklist he knew by habit, then made sure his hoses were out of the way and started spraying in the hard-to-get areas first. He shot seams, edges and mating joints before going to the easier areas.

At this stage he was just shooting on a lighter tack

coat, but the car already began to look beautiful. He then stepped out of the booth and let the tack coat cure for about 20 minutes. Bruce said that one common mistake novice painters make is to hurry the curing of the tack coat. (I must admit, he let it cure for longer than I was used to.) Once the paint sets up to the point of being sticky but doesn't string up when you pull your finger away, you're ready to shoot on the full, wet, final coats.

Bruce stepped back into the booth, tested his gun, then started shooting the hard-to-get areas again. After that he began spraying on the right side in long, even passes, moving at a consistent pace, holding the gun (a conventional high-pressure type) about a hand width from the surface and perpendicular to the panel at all times. He overlapped passes by about 50 percent, and didn't let up on the trigger between passes. The car started to glow and sparkle.

After the booth was cleared of fumes, I was allowed back in to see the result. It was stunning! In the old days of lacquer we used to mix up a final coat of paint that was mostly thinner and let it melt into the coats below to smooth the finish. Modern painters don't have that option, so they really have to be very skillful to avoid orange peel and ripples.

In this instance, thanks to Bruce's skill and experience, the finish was nearly perfect, and I knew that in a day or two as the paint cured more, the finish would become dazzling. After that all the old Jaguar would need to be a knockout is a bit of color sanding and buffing.

Tom's Tack-and-Whack System

For the last 15 years, many car bodies have been made of high-carbon steel that is thin, light and difficult to work when dented. As a result, collision repairs are often done by removing the damaged panels and replacing them with new ones. The only real work involved has been the sanding and prep of the new panel to get it ready for paint. Now, thanks to Spies Hecker, even that is no big deal. With a couple of tacks and whacks, a car can be back on the road the next day. Here's how it's done:

A customer of Tom Horvath's was in a tangle that banged up the passenger-side door and front fender of his VW Rabbit to the point where repairs were impossible. Tom ordered suitable factory replacement panels from a local supplier and they came, as usual, painted in the standard semi-gloss, black, E-coat epoxy primer. (Aftermarket replacement panels are sometimes only shot with a lacquer primer, which is not adequate for Spies Hecker's new process, so wipe a rag with a little lacquer thinner over a small area on the part to verify what's on it. If the primer comes off, remove it all using chemical stripper before priming and painting.)

Painting E-coated replacement panels couldn't be simpler. Start by wiping down the parts with a little degreaser to get any grease, oil or other contaminants off of them. In the process you may find a few small dings that need filling with glazing compound. Use only compound made to be applied to painted surfaces, and only use glaze to fix small imperfections. Sand these spots very lightly when the compound cures. But DO NOT sand the entire panel.

In fact the Spies Hecker warranty will be voided if you sand the panel for this process. That's because their new acrylic urethane primer is specifically designed to bond with the smooth primer already on the panel. Amazingly, the molecular structure of Spies Hecker's new primer is made to bond with the molecular structure of the smooth primer, and sanding disturbs this process.

Tom mixed one part Spies Hecker Permasolid 2K Very High Solid hardener 3315 to one part VHS Wet on Wet surfacer 5190. No reducer is needed or even allowed. Horvath stirred the mixture thoroughly, then poured it through a filter into his DeVilbiss GT1 gravity-feed spray gun. (Tom uses DeVilbiss spray equipment exclusively, and prefers gravity-feed guns.)

As with any of the new paint systems, be very careful to get the mixture right, and make sure you will be painting in a dry environment with an ambient temperature of 65 degrees Fahrenheit or above. If the temperature is too low, the paint will not cure properly. Be sure to wear proper protective gear while shooting because these paints are as toxic as other modern finishes.

Tack and Whack—Shoot on a tack coat and give it about five minutes to get sticky. Shoot difficult areas such as doorjambs first, then shoot the broad areas. Don't let up between passes because you don't want the spray gun to clog. When the tack coat becomes tacky, shoot on a full, wet coat of the primer, again starting with the edges and difficult areas and finishing with the flat areas.

Let the primer dry for at least half an hour. Now mix up the color base coat and shoot it on using the tack and whack (just five minutes between coats) method outlined above. You may want to shoot on two wet coats after the tack coat to get full coverage, but don't overdo it. If paint is shot on too thick it will be prone to problems later.

Let the color coat dry for half an hour, then shoot on the clear using the same method outlined above. Assuming you have practiced a little with your spray equipment before shooting the paint, your new finish should look shiny and beautiful, with only a little orange peel. (Most of the orange peel will be pulled out as the parts cure and the paint shrinks.)

Finally, after the components have cured for a day or so and off-gassed completely, lightly color sand everything with 2000-grit, then polish using a buffer or orbital sander with a foam pad, and some CSI 62-201 Cut 'N Polish.

1. Spies Hecker's new, two-part acrylic urethane primers are designed to be shot directly onto E-coated replacement panels with no sanding beforehand.

2. This VW Rabbit panel and the door behind it were damaged beyond repair, so factory replacement panels were ordered.

3. Clean the panel thoroughly with a quality degreaser and prep solution, but don't scuff it at all.

4. Carefully combine the acrylic urethane primer according to instructions and stir thoroughly.

continued on next page

continued from previous page

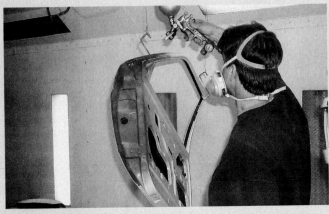

5. First, shoot a thin, misty tack coat on difficult areas, such as doorjambs; then shoot broad areas.

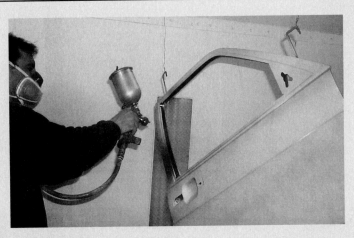

6. Let the tack coat dry so it will get suitably tacky and help the next full, wet coat stick. When dry, shoot on a full, wet coat of the high solid primer and let it dry for at least 30 minutes.

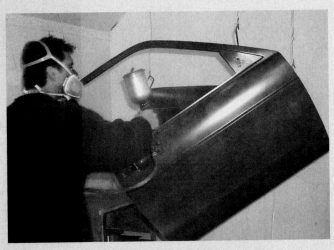

7. Again, use Tom's "tack-and-whack" method when shooting on the color base coat.

8. Let the tack coat get sticky before shooting on a full, wet coat of color.

9. Tom shoots on a wet coat of the color base, which looks sort of dull without the clear.

10. Tom's tack-and-whack system is used again for the clear coat. The parts look super, and the whole process takes only a couple of hours.

Chapter 23

Blending Two-Stage Urethanes

A good safety habit to develop whether you are doing mechanical or bodywork is to disconnect the ground cable from the battery first.

A major advantage to good old-fashioned lacquer was that good automotive painters could blend spot repairs to panels beautifully. Problem is, lacquer is banned in many localities today because of its VOC (volatile organic compounds) content. The stuff ruins the ozone layer and pollutes the air. As a result, today most pros use urethanes, which, as we have said many times before, are extremely toxic, but they don't damage the atmosphere as much.

In many ways the new urethanes are miracle paints. They are extremely tough, durable, and can be buffed out to look magnificent. However, the single-stage urethanes suffer from the same drawback as the old acrylic enamels in that they cannot be easily blended when doing repairs. With single-stage urethanes you have to paint whole panels or find areas where an imperfect blend won't be noticed in order to do spot repairs on them.

The good news is, the new base-coat clear-coat, two-stage urethanes can be blended beautifully if you know how to go it. To find out how, I once again turned to my good friend and co-author Tom Horvath's shop to see the process for myself. As it turned out, he had an otherwise impeccable Ferrari Daytona done in a base and clear coat urethane that needed a couple of small fixes.

The first thing Clemente—Tom's assistant—did was to disconnect the battery. This may seem unnecessary to some, but it is a good safety habit to develop. Any time you work on a car you run the risk of shorting across components and

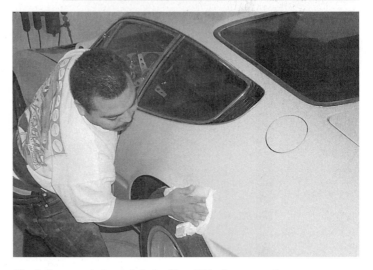

Wash the area to be painted with a little degreaser to remove any fingerprints or oil and grease.

damaging them, or worse yet, causing a fire. And if the car is going to be sitting idle for a while, disconnecting the battery will help keep it from discharging. Next, wipe the car down with a good de-greaser and wax remover like DuPont's Pre-Kleeno.

To begin the repair, first fix any dents or deformities in the panel. Our Ferrari's problem was only skin deep and didn't affect the underlying sheet metal, so all we needed to do was

Our Ferrari Daytona only had a scratch so all we had to do was feather the finish with 500-grit sandpaper.

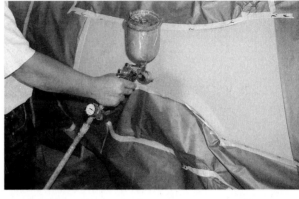

Shoot a little primer on the repair spot. Use only primer recommended by the paint system you will be employing.

Only sand the small area where the repair is to be made, but use a gray Scotch Brite pad to develop a tooth and knock the shine off the surrounding area where the blend is to take place.

Sand the immediate area of the repair using 500-grit sandpaper.

Double tape around the repair so as not to have a hard line where the repair ends.

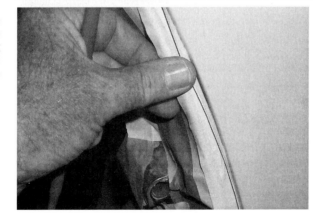

to re-repair your repair because you weren't careful the first time.

Mask around the panel, double taping as you do so you protect the car from overspray. A little extra effort in the masking stage can save some big problems later. Mask out well beyond the area to be painted, and make sure wheels and suspension are covered or masked too. Even today's HVLP paint guns can spit a fair amount of sticky overspray into the air which will then settle on other areas of the car.

Shoot on a little of the primer recommended for your paint system. For instance, if you are using PPG paint, make sure you use their recommended primers as well. As we have said again and again, mixing brands of paint can sometimes prove disastrous due to chemical incompatability. Just shoot on enough primer to cover the bare metal spot and overlap a little bit into the surrounding area.

When the primer has cured the proper amount of time as specified by the manufacturer, wet sand the area using a bucket of clean water, a sanding block,

sand the area flat. You only need to sand in the immediate area where the blemish is, using 500-grit wet-and-dry paper and a little water. Feather the edges of the damaged paint carefully. When you have the area perfectly flat, take a gray Scotch Brite pad and scuff the shine off of the rest of the panel out past where you want your blend line to be by a foot or so.

Next, wipe the entire panel down again with a degreaser to remove any vestiges of wax, fingerprint oils or other contaminants. Finally, use a tack rag to pick up any dust or dirt that might have settled on the area to be repaired. There is no sense in having

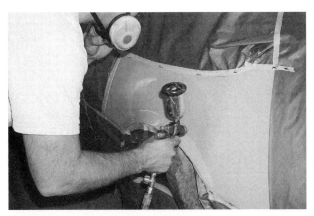

Spray the base color coat only on the repair and work from the perimeter in toward the center.

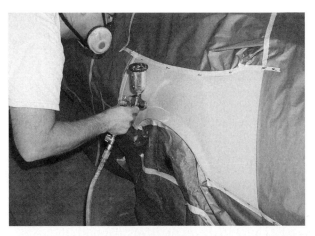

After the recommended cure time for the base coat, shoot on the clear, again working from the edges in toward the center.

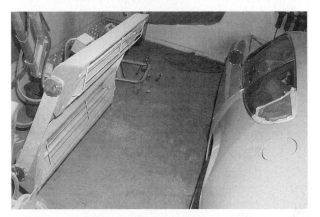

Next, the finish must be completely cured and free of solvents. A heat lamp is the way to go.

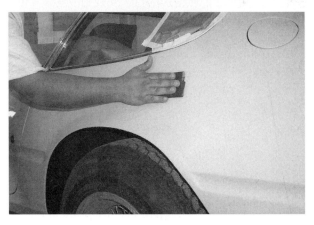

Color sand the final repair and surrounding area using 2,000-grit microfine paper and water.

and 500-grit wet and dry sandpaper. As soon as the immediate area surrounding the dent is perfectly flat with no blemishes or other problems, you are ready to shoot on the color base coat. Just as with the primer, you only need to paint the repair, not the whole panel.

Stir the paint thoroughly together in the proportions required and pour it into a touch up gun. Wear a least a respirator with fresh canisters—or better—a fresh air pack, and shoot on the base coat working from the edges into the center of the repair. You want the paint to be thickest at the center and thinner at the edges.

When the base coat has hardened the proper amount of time, shoot on the clear coat in the same manner as you did the base coat, working from the edges in toward the center. You only need to apply clear to the area that you sanded, but not the larger area that you scuffed with the Scotch Brite pad.

The next step is critical to producing a durable repair. Use a heat lamp to cure the area completely so as to get every last bit of solvent out of it (or let it dry for several hours if you have no access to a lamp). This can be done with a large, professional-style lamp or a portable one. Use an infrared heat sensor to make sure the panel is kept at the correct temperature.

Once the panel is completely cured, the rest is easy. Remove the masking paper and clean the work area. Put a piece or two of 2000-grit microfine sandpaper in a bucket of water along with a couple of drops of liquid detergent and let it set for twenty minutes so the edges of the sandpaper can soften. Go over the entire repair with a tack rag one more time to remove any grit or dirt.

Wrap a piece of the sandpaper around a flexible sanding block. Wet the surface of the entire repair area with water and begin sanding. Work in short crisscross strokes so as to minimize the possibility of scratches. Don't apply a lot of pressure. Just let the sandpaper do the work. Keep the area wet at all times. Periodically check your progress with a clean rubber squeegee.

When all traces of orange peel or Scotch Brite scratches are gone and the surface has a dull, consistent sheen, it is time to do the final polishing. I like my pal Tom Horvath's CSI Cut 'N Polish for this task. Other automotive polishing products make three-stage buffing systems that will do a good job, but Cut 'N Polish does the whole task quickly with just one ultra-fine compound.

Put a little compound on the spot to be polished, then—using a sheepskin pad on a variable speed buffer—polish the area of the repair at slow speed. Don't let the buffing pad get dry because you will burn and damage the paint if you do. When you have gone over the whole area with the sheepskin pad, wipe off the excess compound. Your repair will look great, but there will still be a slight spiderwebbing.

Switch to a yellow foam pad and repeat the polishing process. Again, make sure you polish at slow speeds and tip the buffer so you only use the area of the buffing wheel from about 12 o'clock to 9 o'clock to do the work. If you flat buff, the machine is too hard to control. Wipe off the excess, then switch to a fine gray foam pad and, using the same compound, go over the repair and surrounding area one more time.

Your repair should look dazzling at this point and there should be no trace of a blend at all. The secret is that you are not blending the color coat, but are only blending the clear coat into the surrounding area. Finish by going over the panel with a soft cloth and a little more compound, then apply a coat of pure wax with no cleaners in it. No matter from what angle you view the car, or under which kind of light, the repair should be completely invisible. Even black cars can be done this way. Our Daytona is ready to take a few more trophies, and the repair Tom made will last as long as the surrounding finish. This technique sure beats having to paint whole panels or hiding ugly blends.

Apply the polish and buff the entire area on slow speed using a sheepskin pad.

Next, switch to the yellow foam pad and use the same polish to buff further.

The spot repair is now invisible. You'd never know it had been fixed.

Chapter 24

Color Sanding for a Show-Winning Finish

We stripped my 1955 Beauville wagon to bare metal and fixed all the problems before painting.

Things You'll Need
- Liquid detergent
- Clean, plastic bucket
- Tack rags
- Soft rubber sanding pads
- Rubber squeegee
- DeWalt variable-speed power buffer
- CSI 62-201 Cut 'N Polish or other one-step professional polish
- CSI Q-7 wax or good carnauba wax with no cleaners
- CSI Q-7 Detailer
- Clean, microfinishing rags

Did you ever wonder how they achieve those dazzlingly perfect paint jobs you see on show cars? Paint so deep and shiny that you could get lost in it? Paint that has no hint of orange peel, spiderweb scratches or tiny imperfections? Well the secret is color sanding. It takes a lot of time and attention, but almost anyone can do it at home with only a few simple tools plus some microfine sandpaper and a fine polish such as Clear Coat Solutions CSI 62-201 (www.clearcoatsolutions.com). CSI products may be ordered online, and are second to none.

The basic technique has been around for a long time, but in the past, once you had sanded the car with progressively finer sandpaper, you had to go over it three more times with different compounds—the first of which was rather aggressive—in order to achieve the desired effect. But now, with a new polish that has become available, once the car is properly sanded, you can buff it out to perfection easily in one step with little danger to the finish. Here's how we did it on my 1955 Bel Air wagon:

We carefully removed all of the brightwork, took out the windows, and then stripped the car to bare metal using chemical stripper. After that we got rid of any rust, and made the required repairs. Finally, we gave the car several coats of high-build primer, block sanded it, sealed it and then shot it with a quality base coat/clear coat paint system in the

CSI Cut 'N Polish and Q7 wax and detailer are my favorites for buffing out a car. They are available at clearcoatsolutions.com.

original Glacier Blue with Shoreline Beige on top. We made sure we shot on a little extra clear to allow for the color sanding. We then let the car cure for three days before going to work.

After the car was painted we let it cure, then washed it carefully before proceeding.

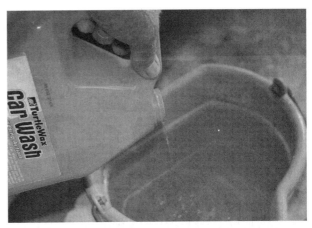

We put a few drops of car wash detergent in the water with the sandpaper to help soften it.

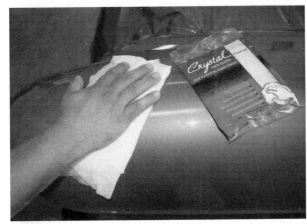

Before beginning color sanding we went over the whole car with a tack rag to pick up grit.

Microfine sandpaper in 1000-, 1500- and 2000-grit is used to color sand the car.

My friend Clemente Chavez did the painting and helped me color sand and buff the car out. At the time he hung his spray gun at a shop owned by Tom Horvath—inventor of CSI Polish—whose front office was filled with trophies from prestigious shows. As a result of Clemente's efforts, we began the process with a paint job straight out of the gun that was better than most shops do. In fact, it wasn't easy to begin sanding on that nearly perfect finish, but nearly perfect was not good enough to take home the gold. We wanted no less than show winning paintwork.

Of course, this process requires that there be enough paint on the car to safely color sand it. You can't do it to new cars because their finishes are only three to four mils thick, so you would ruin the UV protection of the clear coat in a hurry if you tried to color sand it. But a good respray worthy of a classic will be between 13 and 25 mils thick, thus allowing us to sand out any irregularities.

But just to be on the safe side we used a magnetic paint gauge to make sure we had enough paint on the car. The way these gauges work is, the higher the pull-up handle goes before the magnet becomes unstuck, the thinner the paint. Thicker paint insulates the car's sheet metal from the magnet inside the gauge, allowing it to pop off more easily.

To begin with, we washed the car thoroughly with a little detergent and clean water to get rid of any dust or grit that could make deep scratches in the paint. Then, just to make sure the car was grit-free, we went over it again with a tack rag. These rags are impregnated with sticky varnish that will pick up any contaminants. You only need to pass them lightly over the paint to remove dirt.

Next, we put a few drops of car wash detergent in a bucket of clean water, and then threw in several sheets of 1,000-grit, microfine sandpaper. We let this stand for about twenty minutes to soften the edges of the sandpaper, and then we dialed in our favorite golden oldies station and got to work.

I highly recommend that you put on the music of your choice before you start, just to help get you in the right state of mind and to pass the time. Color sanding is a Zen-like meditative process in which the doing of each step has to be as important as the result you want to achieve. That is because if you become impatient or sloppy you will not end up with the desired result, and you could even damage your classic's finish. Each step must be done with equal attention, whether it is sanding the hood or

Let it set for twenty minutes, then wrap the sandpaper around a pad to begin sanding.

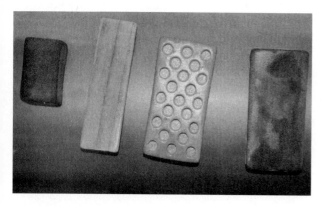

Clemente's sanding pads keep him from making grooves in the paint.

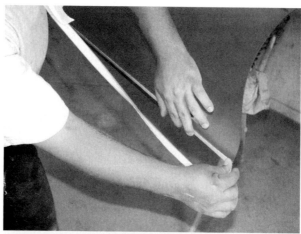

If you are a novice, it is best to tape off edges so they won't get sanded or buffed.

Notice that flat surface of door has been color sanded, but edge is left completely alone.

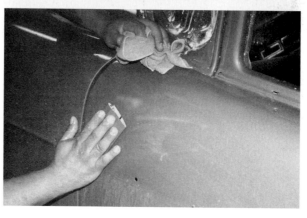

We used plenty of water squeezed from sopping wet rags to keep surface wet while working.

carefully going around window frames and drip moldings.

Clemente uses a number of sanding blocks so he can apply even pressure to the surface he is sanding. A standard sanding pad is good for most purposes, but pieces of wooden stir sticks and dowel can work well for tight areas. A length of hose can also be used as a sanding pad for inside curves. But in any case you will want to use some kind of pad for all of your sanding. That's because if you sand with just your bare fingertips you will make grooves in the paint.

Also, you never want to sand sharp edges such as those along hoods and doors because very little paint sticks in those places, so you can break through to primer in a hurry. In fact a novice should tape off such edges to protect them.

We used sopping rags to wet the surface of the paint being sanded at all times. A trickling garden hose is also good for this task. We didn't press hard and just let the sandpaper do the work. When it got tired and stopped cutting, we tossed out the old paper and got a fresh piece.

We worked in short strokes along the body so as to avoid any long scratches from unseen-grit or contamination. Clemente and I knew that we would have at least two days of working together to get the car color sanded right, so we just relaxed, enjoyed the tunes and kept going. We used clean rubber squeegees for checking our progress. We sanded areas of about two feet by two feet until all traces of orange peel were gone before moving on.

When we had gone over the whole car with 1,000-grit sandpaper, we switched to 1,500-grit and

A clean squeegee is used to check progress. Only when there is no hint of orange peel do you move on.

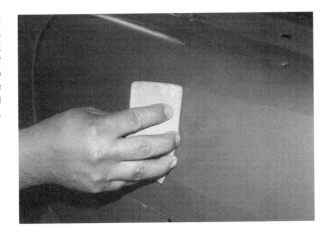

Round dowel is used for curved surfaces, finger is used as a guide to prevent damage.

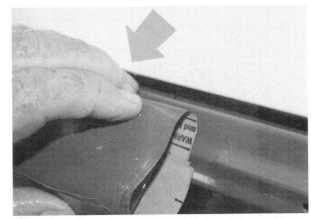

Color sanding takes time and patience. It took us two days to sand my wagon.

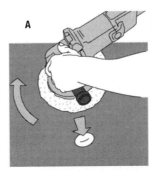

(A) Put the polish on so the material will spin into the buffing pad rather than away and all over the vehicle. (B) Tilt the buffer slightly and buff with the areas between 12:00 and 3:00 or 9:00 rather than trying to buff flat. (C) Tilt the buffer so you polish off of edges rather than onto them so as not to remove too much paint along the edges.

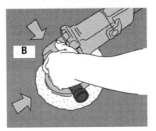
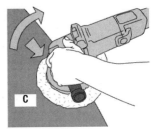

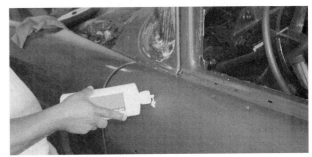

Only a little Cut 'N Polish is necessary for buffing out the surface.

repeated the entire process. The second time around went a little faster because we no longer had orange peel to worry about, but we did need to erase the scratches made by the 1,000-grit sandpaper. Then when we had gone over the whole car with 1,500-grit we started all over again with 2,000-grit microfine sandpaper. And yes, it was a lot of work, but I think the photos nearby show that it was worth the trouble.

When we had finished going over the car for the third time with the 2,000-grit paper and water the car actually had a consistent satin sheen. We then washed it again and went over it with a tack rag one

more time just to prevent any scratches that could be caused by grit that might have settled on the car in the process. At this point even a layer of dust will make scratches if you try to buff the paint with it on there.

Now here's where the real fun begins. We got out our CSI Cut 'N Polish and went to work. CSI Cut 'N Polish is not like most polishes that break down as you use them. Instead it is made of the same superfine material that is used to polish precision optical lenses. Ordinary polishes are like starting out with gravel and grinding it to powder as you buff, and do more harm than good at this level of polishing.

If you have never buffed out a car before, you may want to work with an adjustable speed orbital sander instead of a buffer, just to be safe. It will take longer, but there is very little chance of you damaging the paint on your car if you do. We had done this many times so we used a large, DeWalt variable speed buffer to do our polishing. If you know how to buff properly there is little danger, but it pays to practice a bit before attacking your

Clemente first uses the buffer with the wool pad and keeps moving to avoid damage.

Next comes a soft foam pad to take out the last vestiges of spiderwebbing.

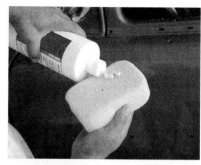

A final wipe-down with a soft sponge gets rid of any haze, before waxing.

The end result is a dazzling, deep, flawless show finish. Now we have to keep it that way.

classic's paint job if you haven't done it before.

The technique we used was to set the buffer on its slowest speed, then shoot on a little of the polish and buff an area about two feet square. The buffer should be held so only the 12:00 to 3:00 or the 9:00 to 12:00 (depending on whether you are right or left handed) positions on the wheel are buffing at any given time. That's because if you buff with the pad flat on the car, you will not be able to control the buffer.

We also made sure we were buffing so the pad rotated away from sharp edges so as not to take any paint off of them. We did not apply pressure to the buffer either, preferring instead to let the pad do the work. A little polish goes a long way, so don't use too much if you decide to try it. If you are spinning wet compound all over everywhere you are definitely using too much. We discovered that a little buffing with a sheepskin pad brings up a gorgeous shine, but if you were to examine the finish very closely in a strong light, you would still see ultra-fine spider webbing.

That's why we went to the sponge pad and buffed the whole car again. By this time the finish was dazzling, but for the final polishing we went over the car by hand with the soft sponge supplied in the kit and a little more of the polish. Then we just stood around and admired our work for a good half an hour or so. The finish was absolutely dazzling! Not a hint of orange peel, scratches or even any surface at all was in evidence. The paint looked as if you could literally reach into it.

What you do with your show-winning finish after you create it depends on how you are going to use the car. If all you are going to do is show it, you only need to go over the finish with a light glaze to protect it. But if you are going to drive the car

regularly, you will want to apply a little CSI Q-7, or straight carnauba wax with no cleaners in it. But apply the wax sparingly because wax can actually build up and dull a finish slightly. The only conceivable reason for applying a thick coat of wax on a finish is to hide scratches, but with this technique and polish, there will be no scratches.

As for washing my 1955 Chevy, I probably won't ever do it. It should never get that dirty. All I should ever need to do is wipe it down with a damp microfinishing cloth and a little CSI Q-7 detailer occasionally. However, if I were going to be driving the car on a regular basis, I will go over it often with Q-7 Detailer to maintain the gloss and help protect the finish.

And if something unfortunate happened such as an unforeseen storm, when I get the car home I will wash it using a only bucket of clean water, a little car wash detergent and a soft cloth. That's because using a running garden hose the conventional way allows water to run down in the car's body and cause rust in areas that are inaccessible.

Chapter 25
Plastic
Bumper Repair

This Corvette has suffered a little bumper damage and a crack thanks to a high curb.

Things You'll Need
- 3M Automix two-part flexible parts repair kit. (There are several types of repair kits for specific plastic bumpers and other parts, so consult your automotive paint supplier to make sure you get the correct one for your application.)
- Primer for plastic parts
- Small orbital sander
- Sanding boards
- Teflon mixing board and plastic filler spreaders
- 80- and 150-grit open coat sandpaper
- 320- and 500-grit wet-and-dry sandpaper
- Masking paper and tape
- Urethane primer
- Touch-up paint gun
- Urethane paint in the correct color (usually base and clear coat)
- Flexing agent

Not many years ago fixing bumpers meant taking them off, straightening them, welding up any rips, then having them replated. It was a lot of work and expense. But times have changed. These days, you can repair bumpers very easily, and that's good because they generally need more repair than the old ones from before the five-mile-an-hour rule. Actually even the metal bumpers of a few years ago were more for looks than for protection and were pretty frail. You have to go all the way back to the cars of the '50s to find bumpers that could plow through a cinder block wall without a scratch.

I was out at Tom's shop the other day when a late model Corvette came in with a bad blemish on its bumper. The break didn't go all the way through, but it was nasty and unsightly looking. Tom Horvath dispatched his master panel beater Clemente to take care of the problem and I got a ringside seat from which to observe. I was surprised at how easily and quickly a repair could be made. Here's how to do it:

If the bumper you need to repair is a single piece of hard plastic, a crack will most likely go right through it. To fix such a crack, remove the bumper, clean it up from behind, then sand it using 150-grit paper. When the area is thoroughly scuffed, apply a strip of fiberglass using the same 3M two-part Automix to attach it. This will provide the necessary reinforcement to chemically weld the crack together.

The newer cars such as our example Corvette have thick foam bumpers with hard shells, and in most cases they can be repaired without even removing them from the car. Start by wiping the area down with a good wax and grease remover. Then sand off all the old paint around the damaged area. A small orbital sander is good for this task, but not absolutely necessary.

If the hard outer shell of the bumper is actually cracked, you will need to vee out the crack to about 1/4" wide and deep using a utility knife so as to create a broader surface to allow filler to bond adequately. The process is much like when you have to patch a crack in a plaster wall. If you don't open it out, the crack will just reappear in short order. When you have the area properly sanded and the crack opened up, finish by washing down the area with a little Pre degreaser just to make sure there is no oil or contamination on the area to be repaired.

Mix together equal amounts of filler (two gobs the size of a dime will do for a start) from tubes A and B just the way you would plastic filler, using a rapid S stroke. And as with any plastic filler don't lift the spreader out of the material while mixing so as to avoid stirring in air bubbles. This stuff starts to set up pretty quickly so work fast, and as soon as the material is a consistent gray with no black or white streaks in it, start spreading it on the crack. The filler is a lot like body filler, but is creamier and goes on quite easily.

Let the filler cure for at least a half hour, then rough it to shape using 80-grit sandpaper. Finish sanding using 150-grit paper. During this process you may notice small holes, edges or low spots. If you do, use more filler to fix these problems before shooting on primer.

Mask around the area to be painted, and try to do it in such a way as to allow you to paint up to the natural lines of the car rather than trying to blend an area. It is extremely difficult for most painters to do a successful blend, so don't try it unless you have no other choice. Also, be sure to mask inside holes and vents so you don't get paint into them either.

Next, shoot on a wet coat of primer especially formulated for plastic parts. Let this cure for an hour or two, then sand with 320-grit. Shoot on two more coats of primer and wet sand with 500-grit. Finally, shoot on the color coats in the matching color and let them cure for the recommended time as specified by the paint manufacturer.

Finish the repair by mixing a little flexing agent into the clear and shooting on the required number of coats. Let the repair cure overnight, then lightly color sand with 2,000-grit sandpaper and a little water so as to blend the surface of the repair into the surrounding area. Your bumper should now look as good as new.

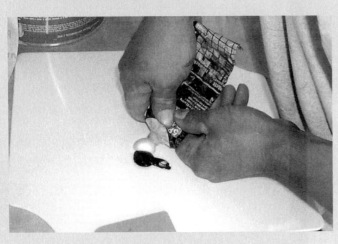

1. Sand off the damaged paint and primer using an orbital sander and 150-grit open sandpaper.

2. Open out any cracks to a vee-shaped contour using a utility knife so the filler will stick.

3. Mix equal amounts of the two-part filler. Two gobs the size of a dime will usually do for a start.

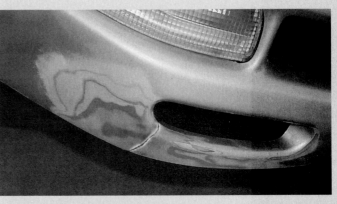

4. Mix rapidly in an S and don't lift the spreader because you will make bubbles if you do.

continued on next page

continued from previous page

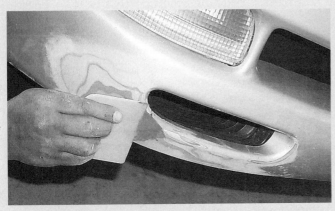

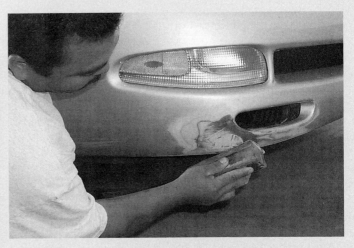

5. Apply the plastic filler just as you would any filler. Lay it on a little thicker than you want it to be when finished so you can sand it down.

6. Let the filler cure for half an hour, then sand and sculpt it to shape using 80- and then 150-grit open coat paper.

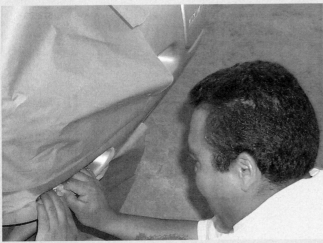

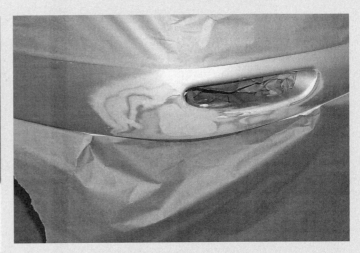

7. Mask off the area so overspray won't ruin the existing paint job. Mask inside holes too.

8. Try to mask the bumper so you have as little feathering to do as possible.

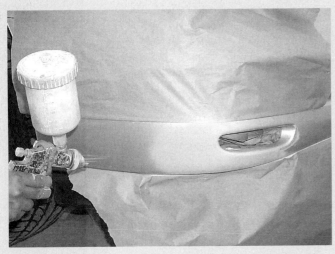

9. Shoot on special plastic primer, then sand and prep for the color coats.

10. Our 'Vette is as good as new, and its bumper looks great.

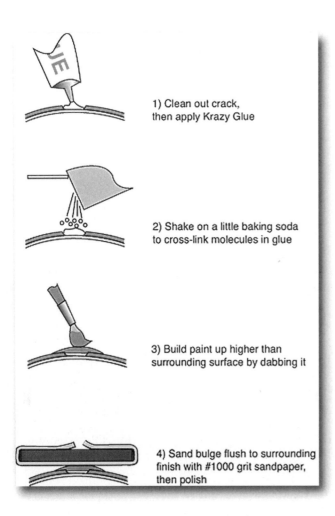

1) Clean out crack, then apply Krazy Glue

2) Shake on a little baking soda to cross-link molecules in glue

3) Build paint up higher than surrounding surface by dabbing it

4) Sand bulge flush to surrounding finish with #1000 grit sandpaper, then polish

Here are the basic steps to fixing cracked paint for good.

Chapter 26

Fixing Paint Cracks

Things You'll Need
- Single-edge razor blade
- Krazy Glue (Cyanoacrylate)
- Baking soda
- Touch-up paint
- Sandpaper 320, 600, 1,500, 2,000 grit
- Polishing compound
- DuPont Pre-Kleeno
- Gray Scotch-Brite abrasive pad

One of the most sublime automobiles ever built—Ferarri's Daytona—is a masterpiece of engineering and style. But even the Daytona isn't perfect. No car is. As many owners of these works of art will tell you, stress cracks can form in the old lacquer paint around the headlight doors and side marker lamps due to vibration and flexing. The usual way to fix such problems is to clean out the cracks and daub paint into them until they are over-full. This is allowed to cure, and then sanded level. The problem with this method of repair is it doesn't last long.

Fresh paint put on top of old paint does not bond well, so cracks reopen in a hurry. Obviously, the very best way to deal with stress crack problems if you are doing a complete restoration is to put in a few reinforcing brackets at critical spots. But if your car's finish is in nice shape you won't want to resort to such radical surgery, nor is it necessary. Here's what to do:

I saw this little trick demonstrated recently and was quite impressed. Newscaster Paul Moyer took his newly acquired black Daytona in to my pal Tom Horvath to have it touched up and detailed, and I happened to be there at the time. Moyer knows that when it comes to fine paintwork, Tom

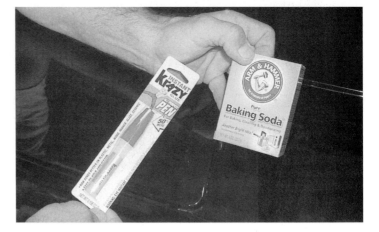

These inexpensive crack repair components are readily available at the supermarket.

Horvath is second to none. He has done Ferraris and other less stellar classics that have taken home the gold at Pebble Beach.

Tom's method for fixing stress cracks surprised me because of its simplicity. All it takes, other than the usual touch-up tools, is a single-edge razor blade, a small tube of Krazy Glue

Use a single-edge razor blade to open and bevel cracks down to bare metal.

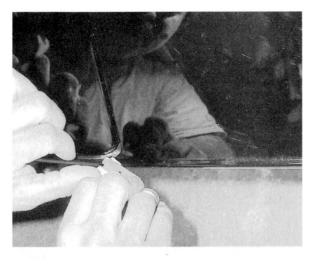

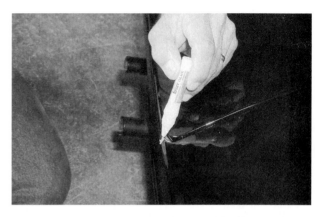

Squeeze a little Krazy Glue into the crack until it is slightly overfilled.

Here is how the crack should look after you open it up. Start by cutting out the crack down to the bare metal using the single-edge razor blade. Bevel the edges of the crack so the glue will have

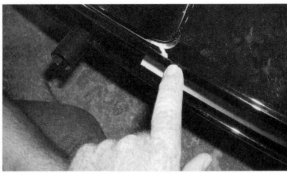

a good surface on which to bond. Make the cuts about 1/8 inch wide. Pop off any flaking paint as well.

Apply the magic baking soda, and poof!, the crack is fixed.

Be sure to mask off adjacent panels as necessary before applying glue.

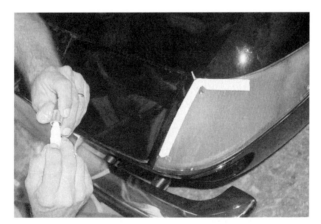

(Cyanoacrylate) and baking soda. You will also need the usual microfine sandpaper in 1,500- and 2,000-grit, as well as a couple of sheets of 320- and 600-grit sandpaper for the rough work. And you will need polishing compound to buff out your repairs and restore your car's finish to its original luster.

Krazy Glue is amazingly strong and will stick almost anything to anything else. A word of caution before we begin though: If you get Krazy Glue on a finger, wash it off immediately with acetone or nail polish remover because if you touch two fingers together, you won't be able to get them apart unless

you use copious amounts of acetone and work your fingers back and forth to slowly loosen them. In fact, dermatologists are now using the stuff to close small cuts, and that is sort of what we are going to do too.

Open the tube of Krazy Glue and squeeze a line of it into the center of the crack to just slightly overfill it. Dribble the baking soda into the Krazy Glue to help it cross-link. Let this set up for a few minutes while you do other cracks. The Krazy Glue will knit the paint back together, making it stronger than it was originally. Brush off the excess baking soda, then use a small piece of 320-grit sandpaper to sand the repair flat with the surrounding paint.

Wash down the area around the new repair with a little DuPont Pre-Kleeno to get off any grease or wax. Mask off any adjacent panels such as headlight doors, that you don't want to be included in the paint blend, then go over the area surrounding the repair for a few inches in all directions using a gray Scotch Brite abrasive pad, which is the equivalent of 320-grit sandpaper.

At this point you will need to acquire a small

After the glue hardens, use a gray Scotch-Brite pad to scuff the paint around the repair.

Mix matching paint according to the manufacturer's instructions.

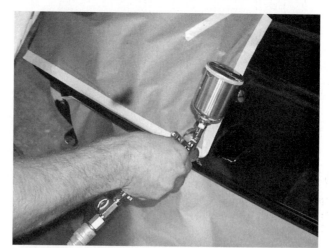

Use a touch-up gun to shoot on the paint, but leave at least 3 inches of scuffed original paint around the repair.

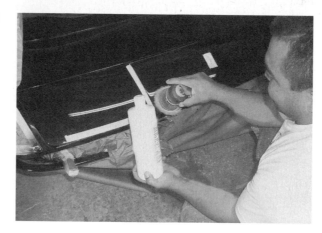

Use a professional grade of polish such as CSI Cut 'N Polish to polish out the entire area.

amount of the correct paint to match the car. In the case of Paul Moyer's Daytona, all we needed was a little black lacquer to do the job. If your car were painted in the new urethane base coat/clear coat systems, you would need both the base and clear, but the blending technique is the same. However, if your car has been repainted in acrylic enamel or single-stage urethane, you will need to scuff the entire panel to the edges and repaint it, because a decent looking blend with such paints is nearly impossible.

Mix your paint according to the manufacturer's instructions, and put a little into a touch-up spray gun. Shoot the paint on over the crack, working from the outside in toward the middle of the repair. Leave at least an inch of scuffed paint all around the repair so you can color sand and blend it with the surrounding paint.

Let the paint dry for a day or two, then lightly sand the area with 1000-, 1,500-, then 2,000-grit microfine sandpaper wrapped around a rubber sanding pad dipped in water. Next, put a little buffing compound on a sheepskin buffing pad and buff the entire scuffed area with the buffer running on the slowest setting. Go to a gray foam buffing pad and repeat the procedure, going out beyond where you buffed before. Finally, go a soft yellow sponge pad and buff once more. Finish with a soft cloth and a little more polish. Your repair will now look faultless and it will hold up for years.

Chapter 27
Body Panel Assembly Tips

Things You'll Need
- Old blankets
- Small carpet squares or pieces of plastic tubing
- Bits of plastic foam
- Measuring tape
- Original fasteners, cadmium plated, painted or cleaned as required

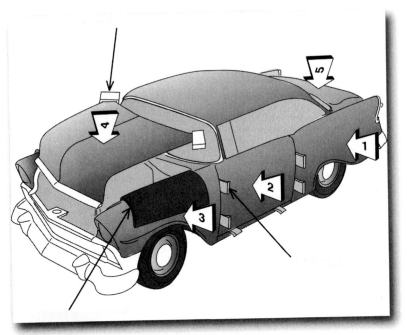

Here is the sequence for assembling most cars. Use carpet patches, blankets and bits of plastic foam to protect parts.

Putting a mass-produced car back together isn't as hard as it sounds. That's because most of the engineering lavished on it in the first place was to make it easy enough to construct on a moving assembly line that it could be done by people with no previous experience. The idea was to break the building process down into simple tasks that could be learned in an hour or two. Then the assembler only had to do one simple thing hour after hour. Before long he/she got very good at it.

I know this because I worked on a Ford assembly line back in the '60s hanging front fenders on Falcons and Fairlanes. Even after forty years I remember every step of my job vividly. I started laying in a piece of strip caulk along the inner fender flange. After that I pushed a piece of plush carpeting against the front edge of the door to protect the fender and the door from dings, and to get the gap right between them. I then picked up the front fender and located it on the inner fender, lined up the boltholes with an awl, and buzzed in the top bolts. The final step was to reach down and slap the rocker panel in place so my partner in the pit could install the bottom bolts.

The job was so simple that we could install almost a fender a minute for eight hours a day with very few problems. But if you had said to me back then "now we want you to assemble the entire car," I would have been at a loss. At that

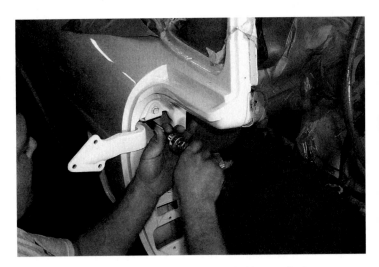

Begin hanging doors by first installing their hinges. Just snug up their bolts so you can make adjustments later.

time I hadn't considered that the whole process was just a sequence of simple tasks, and that yes, given enough time, I could put a car together without much drama. I have since built several cars from the ground up, so I know it's true.

Hinges mount differently on different makes. On this '57 Thunderbird they slide in through slots. Others may fold against the door.

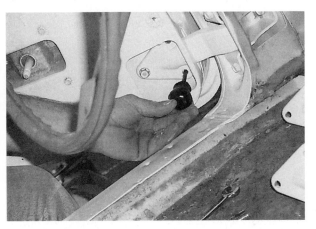

Use an awl to align bolt holes in hinges while an assistant holds the door in place.

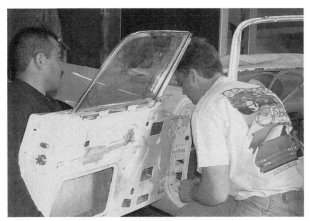

Tom Horvath tightens the bolts while his assistant, Clemente, supports the door.

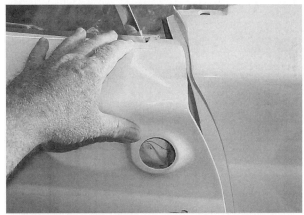

This door hangs way too low at the back. To fix, loosen the hinge inside the door and lift up on the door's rear edge a little.

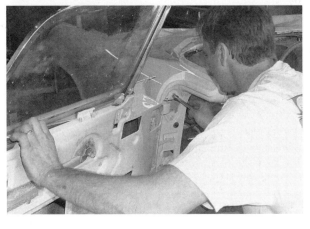

Next, we discover that the door is just a little too far in at the top, so Tom loosens the hinge at the cowl and eases it down.

Together Again

Before attempting to assemble your car, let its paint cure for the recommended time specified by the manufacturer. Otherwise you run the risk of such things as masking tape—applied to prevent chips along edges—actually pulling the paint off the part when you go to remove it. A partially cured paint surface is soft and easily damaged. And fixing problems other than small chips means repainting the panel.

Also, when assembling cars, don't wear jeans with metal rivets on them, and if you have to wear a belt be sure to wear the buckle on your hip, not under your belly. This is not a fashion statement. It's just intended to prevent unwittingly scratching a panel while fitting it. I would also remove any medallions or rings that could damage the finish too.

Align the Body

Assembling the body of a car starts at the rear and works forward. The rear quarter panels and trunk

area on most cars is rigid so it does not move out of alignment with the frame unless you take the body off of it. If you did take the body off the frame, you will need to put back any shims you removed, and replace any body cushions or rubber pads.

On most more recent cars the body lines up on a hole in the frame that corresponds to one in the body up near the transmission. You can hold the body in alignment using a pin at this point while you bolt the body in place. Take multiple measurements as you go though, to see that the body is properly aligned on the frame front-to-back

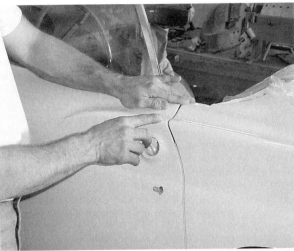

A newly hung door should be just the tiniest bit high at the rear to compensate for wear and tear, but you should not see the difference at a glance.

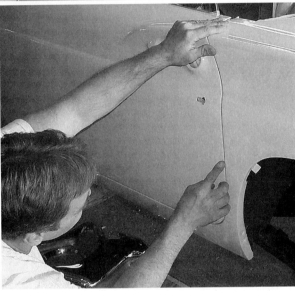

Tom carefully checks the gap all the way around for consistency and to make sure the door isn't in or out at any corner.

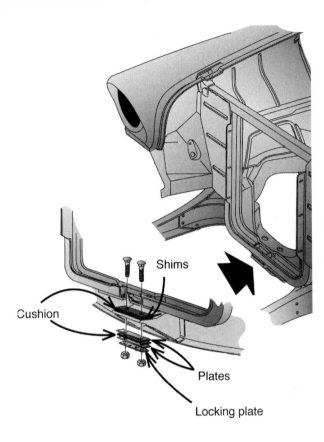

Shims

Cushion

Plates

Locking plate

When installing the radiator cradle and bulkhead, be sure to put any shims back where they belong, and replace any pads or cushions.

and up-and-down at each corner.

Once the body is attached it is time to install the doors. When they are lined up properly they will provide a datum line for front fender alignment. Most of the time, hinges and bolts are painted the same color as the body of the car. To paint bolts, push them through holes in a cardboard shield to mask the threads, then paint all of the hinge bolts at once. Let them dry thoroughly before installation.

Install the Doors

To hang a door, install the hinges in their pockets in the body first. Often hinges are unique as to left and right and top and bottom so don't mix them up. Use an awl to line up their holes and then just snug up the bolts for now. (You will need to loosen and adjust them later, so don't over tighten them.)

Next, get an assistant to hold the door while you slide the hinges into place in it and put in their bolts. Again, just snug them up. Use carpet scraps

to protect the door and body until you get things adjusted close to where they need to be. Now slowly ease the door closed, being very vigilant to make sure the door isn't going to bind anywhere. Doors—whether on cars or houses—must be plumb on two different axes. That means they can't hang down low at the back end, or too high at the front, and it also means that they can't stick out at the top or bottom either.

It can easily take 7–10 small adjustments even for pros to get a door so that it is flat to the body and parallel to the sill, so take your time and be patient. Have your assistant make small moves while you loosen and tighten bolts and inspect your progress. Use body belt lines as reference lines to double check yourself. When the task is completed, the door should move with complete ease, and fit perfectly flush all around with an equal gap.

It should also be just the tiniest bit high at the back, but only so you can feel it with your finger—not so the extra height is visible at a glance. This is to compensate for hinge pin wear and a certain amount of sag caused by people using the door to brace themselves while getting in and out of the vehicle. Once again, if the door is a little high or low at the back or front when compared along a belt line or crown edge, you will want to readjust the door.

Front Fenders

Install the radiator cradle and bulkhead, then install the inner fender panels to it. The radiator cradle and bulkhead needs to be vertical, and it must be at right angles to the frame rails from left to right. If one side of the radiator cradle is a bit forward, the other will be a bit too far back and the fender on that side will jamb up against the door. Once you have the radiator cradle and inner fenders aligned the best you can, just snug up their bolts.

Install necessary caulk or fender insulation. Use scraps of carpeting at the door's edge to maintain the correct gap and avoid nicks. With the fender sitting in place, make sure its gap with the door is consistent and correct before installing bolts. Now check to see that the fender isn't pointing up or down at the front, and whether it is pushed in toward the center of the car, or out away from the center. Make necessary adjustments in small moves.

To get a fender to line up you may need to loosen not only its bolts, but those of the inner fender and radiator cradle as well. Use an awl to line up the bolts at the kick panel and install them. When everything is lined up properly, tighten all of the front end bolts evenly, checking your progress to make sure you haven't pulled or pushed anything out of alignment.

Hood

Depending on the car, it can sometimes be advantageous to hang the hood before adding the fenders. My little '67 Morris Minor has external hood hinges that you install as far forward as they will go initially. Next you install the grille and headlight assembly. Then you attach the hood and adjust the gap between hood and cowl. This generally means moving the hood back slightly. Once the hood is in place and lined up, you then add the fenders and adjust them out or in, up or down.

On the other hand, most modern American cars require that the front end be on the car before the hood is hung. If you scribed around the hinges before removing them and the hood you should not have much trouble installing the hood. But if you did not, or there was collision damage that caused the front end to be misaligned, you may have to start from the beginning. In that case, install the hinges up as high on the cowl as possible and with their notched cams rotated back as far as possible. Snug their bolts up tight enough so they won't slip down during hood installation.

Sight along belt lines to make sure alignment is right. This works for both doors and front fenders.

Attach the hood to the hinges as far forward as it will go and snug up the attaching bolts. Slowly lower the hood, watching for problems as you go. Chances are it will be too high and too far forward. It may also be angled off to one side or another in relationship to the fenders. Even if the rear edge of the hood aligns with the cowl and has a consistent correct gap all the way across, the fenders may have to be shifted slightly to get the gap between them and the hood right.

Loosen their attaching bolts and push them into alignment. If, when you bring the hood down the opening between the fenders is too narrow, you may have to use washers as shims in the fenders to open the gap. It is also quite possible to have the whole front end slightly cockeyed, so measure, think things out, and make small adjustments until you have fixed what is wrong.

Next, adjust the hood, front to back. If it is too far forward, use pieces of rubber tubing or a couple of scraps of carpeting to get the gap right between the hood and cowl. Once the hood is lined up so it has the correct gap all around, you need to align it as to height. If the hinges sit up so the hood is too high in back, loosen their bolts and tap the hood down a little, but make sure you go down the same distance on each side.

When you have the hood fitted in the rear, check out its level as it comes forward. If it is too high or too low you can adjust it by turning the little rubber buttons attached along the fenders that position the hood. Also, later if you experience hood flutter or wobble on the road, these are the items you will need to adjust to eliminate the problem.

Trunk lids are installed the same way as hoods, but aren't generally as difficult to position. Take your time, ease the lid down slowly and don't let it bind. Make small adjustments at any one time, and when the deck lid lies flat and gap is consistent all the way around, you are nearly home free.

Latches for hoods, trunk lids, and doors all need to be installed and adjusted too. Most of the time they can be moved in or out, up or down to accommodate one another. Smear a little white grease on one side of the latch and close it carefully until it makes contact with its other half. Note where the grease makes contact and adjust accordingly. Doors should close easily with a reassuring "chunk" sound and should not hang up at all on their latches.

Hood and deck lid latches need to be lined up carefully so they fasten properly and open easily. If a hood latch is too low, adjust the little rubber buttons along the fenders down a little. If the hood is too loose, it will wobble and flutter and possibly pop open. Also, if the hood latch is misaligned, the latch may not come open easily.

Measure carefully and be patient. Keep adjusting until you have it right. Just remember that this is not that hard to do. A few years ago a couple of people hung that same panel correctly in about one minute. When you are learning, the process takes longer, but there is no magic to it, nor does it take any fancy tools.

Hood on this early postwar Buick is hung first, then fenders fitted, but most later cars are done the other way around—fenders first, then the hood.

Complex layers of flames in striking colors really wake up this classic Bird's appearance.

Chapter 28
Kustom Paint Tips & Tricks

I once helped a friend flame his '50 Ford when we were in high school. The whole car had been shot with black primer, and he had never gotten around to shooting on the color coats, but he decided he had to have flames anyway. We whipped out a piece of chalk and drew on the flames (I did the drawing because I was sort of the artistic type) then we covered up where we didn't want paint to spatter using masking tape and newspaper. We then shot on a lot of bright red up front using aerosol spray cans. After that we sprayed some yellow and feathered it into the red to make the tips of the flames.

The results of our efforts were uh . . . marginal, but the old Ford no longer looked like a gigantic barbecue briquette. Instead, it looked like a gigantic briquette with flames crudely painted on it. At that time, back in the '50s, people had been painting crude flames on hot rods for years, but a creative maniac by the name Von Dutch—along with Big Daddy Roth and a few others—began turning custom, or "Kustom," car painting into an art form.

Today, the process has evolved into a major industry with superstars like Craig Fraser doing gorgeous surreal flames, skulls, lightning flashes and motion streaks in unpredictable combinations that really knock you out of your shoes when you first see them. And the people who are really talented at this new art form are busy making lots of money doing it too.

If you want to try your hand at Kustom painting we can get you started here, but if you are serious about learning the techniques, I suggest you pick up a new book by Craig Fraser called *Automotive Cheap Tricks & Special F/X* published by Airbrush Action, Inc. In it, Craig goes into how to achieve all kinds of effects, and what kinds of equipment you need to do it. Nevertheless, you will have to put in some time practicing, and you will have to expand your sense of creativity and playfulness to really do the killer stuff.

That is because art is a way of seeing things, and God is in the details. These are quotes from famous artists and designers who said it a long time ago and a lot better than I could. To do this kind of painting you need to really analyze the effects you want to reproduce and figure out how they were done and how they can be rendered with an airbrush and paint. You must also develop a facility with the tools, as well as habits of meticulous attention to detail in order to bring off the effects properly.

Tools of the Trade

Various kinds of paint have been used over the years for Kustom painting, but today it is almost all done in urethane base and clear coats. The compatibility, durability and beauty of the urethanes have made other types of paints

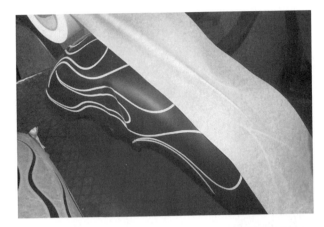

After drawing the flames in vinyl tape we spread on wide masking paper.

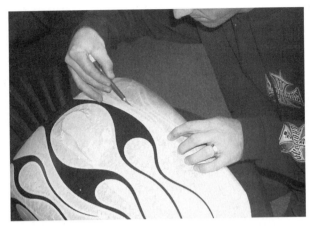

The next step is to carefully cut through the masking paper down to—but not through—the vinyl tape.

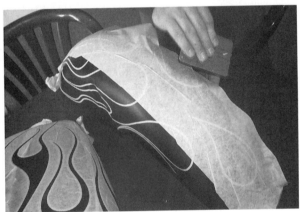

A Bondo spreader is great for smoothing out wrinkles and getting the paper to stick.

Peel the masking paper back on itself slowly to expose the areas to be flamed.

obsolete for Kustom painting. Besides, House of Kolor makes just about every color you could ever want in urethane, and the quality is superb. And to make the whole process easier, complete Craig Fraser House of Kolor starter kits with everything you need except the airbrush are available from Coast Airbrush in Anaheim, California.

How It's Done

Let's start with a relatively simple flame job on a motorcycle fuel tank. Our example tank was primed and painted with black urethane. Then the tank was scuffed with a red Scotchbrite pad and water to give the painted surface tooth for the next layers of paint to bond to. After that we used 1/8" wide vinyl tape (which is easier to work with than masking tape and leaves a nicer edge) to create a flame pattern.

We followed this with a couple of sheets of special masking paper that we massaged into place using a plastic Bondo spreader. This is especially convenient stuff because it covers large areas quickly, and it is transparent enough to see through for cutting. However, 2" wide masking tape will do the same job, albeit with a little more effort and care.

Once the masking paper was in place, we took out an X-acto knife and installed a fresh blade. We

then cut carefully through the adhesive masking paper, but not through the 1/8" tape underneath. This is important because a cut through the paint layer underneath could result in lifting and problems once the clear coat is applied. If you do cut through, you will need to make a spot repair before going further.

Next, we carefully peeled the tape back over itself to prevent it sticking and tearing. Once the masking was removed we then put a little cleaner on a rag and carefully washed the exposed areas to make sure there was no tape residue or finger oil on the finish. We then mixed the base color, thinned it for use in an Iwata airbrush and shot on a tack coat.

When that became sticky, we shot on a full wet coat of base color. After that had set up for about half an hour we came back in with light mists of metallic candy color around the edges of the flames. After working all of the flames several times until the modeling was to our liking, we then mixed up some pearl white and created some highlight lines, giving the flames a rounded, sensuous effect.

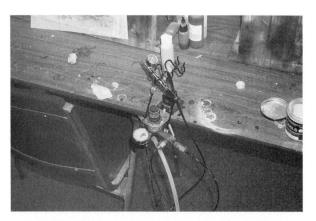

An Iwata airbrush and thinned urethane color coat is used to create the effects after which the whole thing is clear coated.

If you can't draw, stick-on stencils are available to create many of the effects you want.

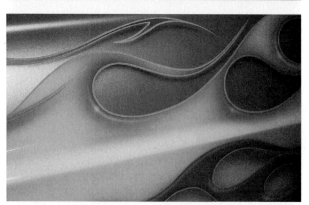

Several passes with an airbrush and thinned metallic toners created these 3D fat flames.

This amazing look was created by abrading aluminum with a sander, then flaming over it. Clear urethane protects the whole thing from dulling.

Dave at Coast Airbrush made the whole process look easy, but that is because of his extensive experience wielding an airbrush. If you decide you want to teach yourself to do Kustom painting at home, you will want a starter kit, a good airbrush, pressure regulator and compressor. Begin your training by drawing flames or other effects on stiff paper or white plastic sign blanks. Mask or use a template to create your pattern, then mix up and thin the paint according to instructions. Then try different pressures and strokes.

Airbrushes are fun to play with, and with a little effort you can learn to control them pretty easily. Try shooting through plastic ellipse templates, circle

templates and other masks to learn their effects. Trace a car or a picture on a plastic sign blank and practice rendering shapes. Don't expect to be good at it right away. You have to practice before the process becomes natural to you.

Pinstriping

This is the most difficult part of the whole process and takes the most practice to learn. Old-time pinstripers like Von Dutch used One Shot Bulletin Enamel for striping because it will cover in one pass, and can be mixed to flow out very well. The old pros—as well as most of the new guys— use a dagger brush that has a long tail that can be pointed and dragged along to make a clean consistent line.

Another tool that can be used for striping that will do a good job is a Beugler Precision Pinstriper. These cut the learning time down by quite a bit for pinstriping because they come with heads that can lay down a perfect width stripe every time. It takes a lot of practice to be able to make a consistent line with a dagger brush. Detachable heads that can make thick and thin parallel lines or two lines of the same width are also available and work well. But no matter how you go about it, plan to put in a few hours mastering the pinstriping technique before attempting to stripe your car.

If you have trouble mastering the art of striping,

Coast Airbrush sells starter kits that contain everything you need except an airbrush and compressor.

The only limits of Kustom painting are the artist's imagination and his mastery of technique.

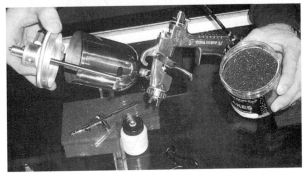

Iwata makes a special airbrush in the $500 range that has a propeller in it to keep even large metal flakes in suspension.

that has a small propeller in it that will keep even large, heavy metal flakes in suspension so your effects will be consistent.

Kandy colors are created by shooting on silver, gold, or bronze metallic paint, and then coming back in with transparent coats of red, green, blue or any other color you would prefer in order to get the effect you want. If you keep shooting on more kandy coats the color will eventually get very dark and subtle, and in the case of red, it will essentially become black. The only limitation to what you can do with this effect is the fact that you don't want the paint buildup to go beyond about 5–7 mils thick. Otherwise your work will be prone to shrinking and cracking when exposed to heat and sunlight.

Extra Special Effects

There seems to be no boundary to the possibilities for unique and creative effects. Aluminum can be polished and scuffed in interesting ways to create highlights that move and shimmer as you walk past them. Flames and other effects can then be put over such surfaces to make the effect more dramatic. The whole thing can then be clear-coated to keep the aluminum from becoming dull.

Artfully torn masking paper can be used to create the effect of an explosion, or ragged torn metal. Wood graining can be simulated to a remarkable degree. Cars have even been made to look as of they were made of granite complete with cracks and fissures. Motorcycle fuel tanks have been done up like cracked eggs, with nightmarish predatory birds breaking out of them. Even shimmering portraits of your favorite rock star or religious icon can—and have been—realistically rendered.

If you are serious about learning more, check out Coast Airbrush's website at www.coastairbrush.com. You will find a gallery of Kustom art, a comprehensive catalog of supplies, and how-to videos, books and tapes. They also offer classes if you happen to live in Southern California.

Once you develop command of the airbrush and a control of the paint medium, the possibilities are only bounded by your imagination. And if you are particularly skillful and talented in the art of Kustom painting, there is lots of money to be made. Little did we know 40 years ago in the era of Von Dutch and Big Daddy Roth that we were only seeing the beginning of an art form that would continue to grow exponentially with no end in sight.

don't fret, because there are people who do nothing but pinstriping who can do a good job of finishing off your Kustom paint work. Check your phone book for stripers and custom painters in your area. The custom painters will know who is hot in your locality. Just make sure that whether you stripe your vehicle or someone else does, a little of the hardener used in your urethane paint is mixed into the striping paint so it won't lift when you clear-coat.

Clear-coating itself can be used to achieve all kinds of smoky or kandy effects. Small amounts of pearl or metallic particles can be mixed into the clear that will provide a shimmering or dazzling feel to your work. Iwata even makes a special spray gun

Chapter 29

Powder-Coating at Home

Eastwood sells powder-coating equipment for the amateur as well as the pro. They also sell infrared lamps and temperature sensors for coating larger items.

Things You'll Need
- Eastwood Hot Coat System kit
- Electric oven (don't use the one you cook with)
- Powder in the colors of your choice
- Paint stripper
- OxiSolv or Evapo-rust rust remover
- Eastwood's Pre for removing oily contaminants
- Particle masks
- Latex gloves
- Heavy work gloves
- Plugs and high temp tape for masking

In the original planning for this book I was going to have a chapter on painting accessories, but that no longer makes sense. I mean, why paint them when you can powder-coat them? Powder-coating is many times tougher than paint, repels solvents, and looks gorgeous. It's a lot easier to apply than paint too. And there is no curing time before you can return the parts to service.

The technology was invented in Europe about 40 years ago as a means of cutting down on VOCs in the air at auto manufacturers' plants. Powder-coating is really tough and looks as good as paint. In fact, it can take a direct blow from a hammer most of the time without cracking. Its toughness isn't because the stuff is so hard though, but because it is resilient and because it bonds to metal far better than paint does.

Until recently the big problem with powder-coating was that you had to take your parts to a pro, pay a hefty price, and hope they didn't lose anything or mess up the job. But now, thanks to some very smart people at the Eastwood Company, you can powder-coat parts at home and do as good a job as the pros can for a fraction the cost.

Eastwood sells two kits and either will do the job very well. I have used both of them and am very pleased with the results. The least expensive kit—intended for the home hobbyist—sells for under two hundred dollars. It is simple to operate and there is nothing to wear out in it, so it will last

for years. A professional unit is also available for more that is really a pleasure to work with and is the kind of tool that can stand up to years of heavy use.

Eastwood's Hot Coat guns may seem a little pricey at first glance, but when you consider what quality paints and primers cost, and the questionable durability of what you have when you are finished, they are a bargain. You could easily run up a $200+ tab having a few items coated at a professional's shop and that's more than the price of the basic kit. Besides, once you own the equipment the powders for the process are your only expense, and they are cheap

Hundreds of colors are available, including all the major manufacturer's engine colors, plus gloss, super gloss, semi-gloss and flat black or white. There are also clears and translucents in every color imaginable that create a candy effect when used on chrome or aluminum. In addition there are high temp coatings in any number of hues, and even textures like wrinkle finish, hammertone, and silver or gold vein through black. I suggest you order an Eastwood catalog. (See the Resources section on page 147) and take a look at the possibilities.)

How the System Works

Here's how the system works: The part to be coated is grounded, then electrostatically charged plastic powder is

The Eastwood Company's powder-coat Pro kit is durable, high quality, and a delight to use.

Once the parts are clean and ready to be powdered, touch them only with latex gloves so as to avoid finger oils which will affect the coating in much the same way they do paint.

If you want to powder-coat professionally you might want to invest in an expandable curing oven.

Use a little Pre or other degreaser to clean any contaminants from your parts after they have been stripped of paint and rust.

solvents, this stuff is virtually impervious. It's on there forever, unless you take copious amounts of paint stripper to it.

The gun used to shoot on the powder puts out a charge of 6,000 volts DC. It could be unpleasant if you touched the tip, but the shock won't really harm you because it is like those jolts you get from crossing cheap carpeting. You won't want to zap yourself if you wear a pacemaker, but the device is no more dangerous than other electric tools.

You will need a supply of compressed air that puts out a consistent 8 to 10 lb of pressure. More than that and you can damage the equipment, so if you have a larger compressor, you will need to acquire a regulator valve unless you go for the pro kit in which case a regulator valve is included. Make sure you attach a water trap such as is used for spray painting too. You can use the inexpensive disposable types if you don't have water traps on the plumbing to your compressor.

For curing components, I purchased an old electric range at a yard sale for $50. (I don't recommend curing parts in a gas oven because of the risk of fire.) I won't be cooking anything in my curing oven either, but my range also came with a working microwave on top that I use to warm up a burrito now and then. When you go out scrounging, try to find a range with an oven timer built in. Otherwise you'll have to borrow your spouse's egg timer. You will also need an inexpensive oven thermometer and these are available at variety stores.

For larger items Eastwood sells an infrared curing lamp and thermometer that can take readings off of your parts for $449.99. That will cure 10" sections of large items at one time and can be moved along until a large part is completely coated. I have one of these and find it quite effective, though you do

blown onto it. The powder clings because of the charge. Then the item is baked in an oven at 400 degrees for about 20 minutes. When the component cools, it looks like it has a coat of enamel on it—but appearances can be deceiving. Unlike paint that is easily scratched, or damaged by

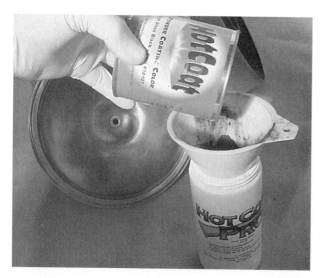

Pour powder into the cup of the gun until the cup is about half full.

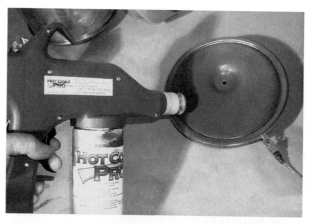

Hold the gun about eight inches from the part and pull the trigger. Work quickly to cover the whole item.

The final step before curing is to remove the little ground lead and give the bare spot a touch-up burst of powder.

need to keep an eye on the temperature while you are working.

Going Professional

The previously mentioned heat lamps work great, but they are a little time consuming if you want to get into powder-coating professionally. In that case you may want to equip yourself with Eastwood's expandable Model 446 Walk-In Oven. As your business grows, you can simply bolt on another module and double its capacity. Without the module, the oven is six feet high by 3'6" by 3'10", which will accommodate most items you would want to coat. These ovens operate on 240 voltage and only cost about $1.80 to operate. But at a cost of $10,000, they are rather expensive unless you are going into business.

Clean Your Components

Prepare parts the same way you would if you were going to paint them. In the case of chassis items, wire brush and media blast them clean. For sheet metal items, use paint stripper to get the old paint off, then use a good derusting agent such as OxiSolv to eliminate any remaining rust. I also like to go over everything with a wire wheel to get any rust out of the pores of the metal and give the coating some tooth to hold on to. Of course I would prepare parts to be painted the same way.

If you are going to be powder-coating a thick, porous, cast-iron or aluminum casting that has been in service, you will need preheat and clean it a couple of times to make sure no oil or grease is left in the pores of the metal. When you have the item down to clean metal, place it in an oven preheated

to 400 degrees for 10 minutes (big parts, such as bell housings take longer) to get the oil and grease in the pores of the metal to bubble to the surface. Then let the part cool and clean it with a good degreaser such as Eastwood's Pre. Repeat this process until no more oil seeps out.

From here on, handle your parts using latex gloves to prevent skin oils from contaminating them. Also, once you've cleaned the part, coat it as soon as possible to avoid corrosion. Unprotected metal can flash over with rust in an hour or two in certain climates. Before shooting on the powder, wrap any bolt threads or areas you don't want coated, such as machined surfaces, with high temp tape (masking tape will do in a pinch but will crinkle and stink when heated). The reason for doing this is if you powder-coat threaded items, they won't fit back together.

Powder Your Parts

Wear a particle mask while working. The powder is not toxic like two-pot paint, but you won't want to breathe the stuff nevertheless. Also, before you powder your parts, fashion some wire hooks from unpainted coat hanger wire, much as you would if you were going to hang a part for painting. You need the hooks to carry parts and suspend them from the oven racks without touching the powder. Next, preheat the oven to 400 degrees Fahrenheit.

While the oven is coming up to temperature, fill the powder cup in the gun about a third to half full of powder and you're ready to start shooting. Clip the ground wire to the part, then depress the trigger and spray on the powder just as you would paint. Hold the gun about eight inches away and shoot quickly because if you linger, a Faraday cage can develop in which electrons will actually repel the powder.

You may experience a situation where the powder tends to build up on raised surfaces, yet avoids recesses. If you do, try to position the part so the crevices are facing upward. That way you will have gravity working for you to pull the powder down in. You can also remove the deflector from the front of the gun and shoot powder directly into crevices that way. And if you mess up the powdering process, simply blow off the powder and start over.

Never touch the tip of the gun to the part because you may damage the gun if you do. When you have an item coated to your satisfaction, undo the ground clip and touch up the bare spot left by the ground wire clip with a short burst of powder.

Pick up your part using needle-nose pliers and hang it from a rack in your oven. The powder clings to the part, but will come off if you touch it. Close the door and "bake" the component for twenty minutes. The powder will become molten and flow out quickly, but it requires the full twenty minutes to cure properly. Use a kitchen timer or oven timer to check the cure time.

When the 20 minutes is up, Let the parts cool for another 20 minutes with the oven off and its door partially open. Your components will now look great, and will stand up to flying stones and strong solvents. You can use this powder-coating tool for air cleaners, intake manifolds, rocker covers, pulleys, master cylinders, battery boxes, suspension parts, brake backing plates and even wheels.

You can also color-sand and buff your parts using automotive compounds. But best of all, you can coat them and put your parts back in place in an hour, unlike painted parts that need to harden for a few days before you can install them.

Twenty minutes later and your parts look as if they have been expertly painted, but the coating is much more durable.

An infrared lamp and temperature sensor can be used to coat parts that won't fit in an oven.

These engine components were done in a variety of colors. The air filter was done in a red translucent that gives it a candy appearance.

Chapter 30

Color Matching

Blue
Jade
Purple
METALLIC ADDED
Red — **Green**
Orange
Chartreuse
Yellow

COLOR WHEEL

The color wheel is what artists use to clarify what will happen when colors are mixed together. Metallic flakes generally make colors lighter.

Matching paint colors is harder than you might think. In fact, it is not really an option for the amateur painter at home. Even many pros can't do it. There are compendious courses offered by automotive paint companies on color and paint matching for professional panel beaters, and there is complex and precise measuring equipment to facilitate the process, but in the end, it still comes down to the human eye. The eye of the beholder. And some of us behold colors better than others.

Such people are worth their weight in gold to the rest of us. I used to go to a renowned fellow named Fernando who worked at a local automotive paint supplier in Long Beach, CA. You had to bring him a piece of your car then wait a couple of weeks, but what he came up with was well worth the $60 he charged to do the match.

For instance, a few years ago I needed to match a metallic light-brown paint on a dashboard, and all I could bring Fernando to work from was one of the gauges. Fernando was able to match the color and the metallic effect perfectly. And yes, I paid less for the pint of paint then I did to have him do the match, but I was happy to do it.

You see, light browns and tans are made up of small amounts of several colors, so they are hard to get right. And metallics are another challenge. The size of the mica particles, the translucence of the color, how wet or dry the

original paint was shot and how it has aged all make matches of such colors next to impossible.

To begin with, the same color seldom matches even from one car to another. That's because during manufacturing of the car, a batch of sea-foam green for example, mixed in September, will be used up and replaced by another batch made by another supplier in October, so they won't even come close to being the same hue. And that's why, when you get those little touch-up bottles of paint at the auto supply they almost never look right when applied.

Also, when you take into account that most colors fade or yellow as they age (a few colors darken) and that all paint oxidizes over time, you realize that there is little likelihood of a can of paint—even though it was mixed to the very same formula—exactly matching the finish on an older car. Even blacks and whites can be a challenge.

Color Basics

Let's start by defining some important terms: HUE refers to the color itself. Purple or green for example. Hues can be primary colors such as red, blue or yellow, or they can be secondary (but still pure) colors which are mixtures of the two colors that are adjacent to them on the color wheel.

For instance, to get orange you would mix red and yellow. To get purple you would mix red and blue. On the other

Before a match can be done, the original paint must be polished and clean.

Tom uses a Spies Hecker computerized system that includes this precise scale. We discovered that a dollar bill weighs one gram. A drop of paint only weighs 2/10 of a gram.

made of, thus complicating matters further. In the end, the only way you can learn how different pigments behave is to mix them together and observe what happens. That's why experience counts a great deal when it comes to matching paint.

If you are shooting metallics, another complicating factor is that the thickness of the layer of paint you are shooting affects its appearance dramatically. A thin layer of metallic paint will be lighter because the mica chips that create the metallic effect will lie flatter and will be closer to the surface.

The first metallic paints were used in the '30s. In those days the mica was ground exceptionally fine and was just used to lighten colors slightly and make the highlights stand out. More modern metallics use coarser mica flakes and are sprayed on dryer and thinner than in the old days.

It is very easy to shoot on a modern metallic finish too thick, which will cause stripes and spots (especially where you overlap for coverage). Thick areas look almost non-metallic next to spots that are thinner so the paint will be shades lighter and more metallic looking in such areas.

Intensity—Another important term to remember is intensity, which refers to the darkness or lightness of a color. It is crucial to adjust intensity before trying to fine tune the hue. As we said before, intensity can be adjusted by adding a complementary color. Clear coats also darken colors, so you need to take that into consideration when doing your mixing too.

Saturation—Yet another useful term is saturation. This refers to the purity of the color. A pure, bright red such as is used on Ferrari's race cars is not diluted or modified by any other color and is therefore saturated. However, if you wanted a less saturated red such as maroon you would add blue and perhaps back off on the yellow. Going the other direction around the color wheel, if you wanted a burnt orange you would add yellow and green to your base red.

At this point we are just discussing basic color theory in order to explain how different colors affect each other, but just knowing the theory doesn't make us capable color matchers. As we said before, that takes an eye, experience, and these days, some expensive equipment if you want to do it right.

With that in mind, I went back down to Tom's Custom Auto Body in Anaheim, CA, to sit at the feet of the master and find out how he does it. Tom's creations have taken trophies at Pebble Beach, and his shop is filled with Aston Martins, Ferraris, Jags and Panteras waiting for his magic touch.

hand, if you throw in the complement of red which is green and opposite it on the color wheel you will kill back, or mute the red making it more neutral looking.

Adding black is one way to darken a color, but that will also kill it in a hurry. Adding white will make a hue more pastel, but will also make it milky looking if you don't add a little complement along with it. An inexpensive way to see firsthand what happens to colors when you mix them is to play around with them using the art software that came on your computer. If you have an application that will let you mix colors and will tell you the percentages, that's even better. You'll be surprised at what results from various combinations.

However, computer colors are just light on a screen, not pigments. Each automotive pigment has its own unique properties depending on what it is

Paint must be stirred thoroughly before a matching test can be done because pigments are heavy and settle out over time.

Catalyst is poured in and the paint is mixed thoroughly again. No reducer is used in modern acrylic urethanes.

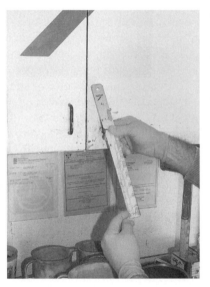

Tom uses a special measuring stick to determine the correct two-to-one ratio of paint to catalyst.

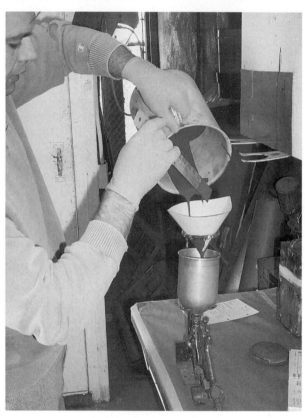

Paint is poured into gun using a proper strainer to filter out particular contaminants.

The Professional Approach

Tom started by carefully cleaning and polishing the panel to be matched. In this case we used a Jag E-Type front clip. You must get down to healthy paint in order to determine its exact hue. As it turned out, the old Jag appeared to be almost a pure red, straight from the red toner can.

Next, Tom got out a can of the paint mixed to the original factory formula on file in his computer. He then stirred the paint for about ten minutes to thoroughly mix the pigments into suspension. This is an extremely important step because many pigments are quite heavy and will lay on the bottom for a long time. Metallic particles are especially heavy, so you must mix metallics even more thoroughly than you would solid colors.

Tom then carefully placed the can of paint on a precise and sensitive scale and noted its weight. (A dollar bill placed on the same scale weighed in at one gram.) The scale needs to be so sensitive

because a drop of paint only weighs in the neighborhood of 2/10ths of a gram and it doesn't take many drops of toner to alter a color considerably.

The paint Tom was mixing was an acrylic urethane, so next he just mixed in the catalyst at a two-to-one ratio and stirred these together thoroughly. No reducer is used with the newer paints due to current VOC (Volatile Organic Compounds) environmental standards. Tom used special aluminum measuring sticks to get his ratio just right.

The next step was to pour the paint into the reservoir on the spray gun using a strainer to make sure no contaminants get into the gun or onto the finish. Tom then went into his spray booth, put on

Horvath uses a special card with reference markings as his test patch.

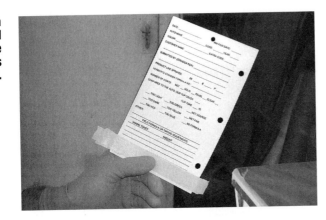

A mist-like tack coat is shot onto the card first in order to avoid orange peel.

Final coats are shot on when tack coat is suitably sticky. You know when you have coverage when you can no longer see printing on card.

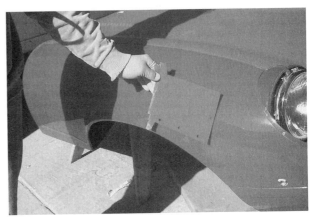

Our factory original color is just a touch orange when compared to the Jag fender.

We let the card dry for a full 30 minutes because automotive paint always dries darker than it appears when still wet. Then he took the Jag fender over to the entrance of his spray booth and put it in nearly direct sunlight. He did that because the color of the light in which you view a hue will affect how it appears. For instance, if you looked at a painted part in late afternoon sun it would look redder than if you viewed it at noon when the light is neutral.

That's because late afternoon as well as early morning sunlight must travel through a lot more atmosphere to get to you. And in Southern California, there is the beige of the smog to account for as well. In fact, the color of the light is so critical that at the paint factories the lights in the matching areas are calibrated at 5,000 degrees Kelvin, which is the color of sunlight at noon. That way they are comparing apples to apples and oranges to oranges, no pun intended.

Another thing that can affect color matching is reflected light. If we park our red Jag next to a black car, the reflected light from it will make the Jag look darker. On the other hand, if a yellow car parked next to our Jag, it could make its red color seem orange. Shades of white can be especially hard to match because it picks up any color reflected in it.

When the color test card was thoroughly dried, Tom placed it on the Jag front fender. The match was close, as you might expect, because the colors were supposed to be the same according to the factory formula. However, it wasn't a perfect match. Our test panel needed just a tiny bit of blue to kill it back slightly because it appeared a little too orange when placed on the fender. Even with the sophisticated machinery, the human eye was the final judge.

To achieve an exact match, a couple of drops of one of the many toners Tom had on hand needed

latex gloves and a mask (he was already wearing long jeans and a long sleeve shirt) and shot a light tack coat onto a specially prepared white card provided by the Spies Hecker paint company.

He told me that the tack coat was necessary in order to prevent orange peel so as to duplicate the finish on the Jag as closely as possible. After letting the tack coat get sticky for a couple of minutes he proceeded to shoot on two or three more coats to get total coverage. Tom knew when he had the paint covered completely (to the point where the white of the card wasn't altering the hue) because the black printing on the card also became hidden at that point.

to be added to our can of paint. The amount was measured precisely using our weighing scale, so we know exactly how much toner we had added. Horvath also jotted down each change he made in a color so he wouldn't forget what he did if a phone call or a customer took him away from his work for a while.

It often takes several adjustments to get a proper match to the existing paint on a car, but it is worth the effort. Nothing looks worse than a slight mismatch between panels. When our paint had been adjusted and the paper test panel dried again, Tom taped it to the fender and it literally disappeared. The match was perfect!

After seeing Tom do a color match in a little over an hour, I gained a great deal of respect for his skill and for the new automated Spies Hecker system, and I gained an even greater appreciation for my old friend Fernando who used to do all of this by trial and error behind his paint store not many years ago. As we said before, it isn't easy. The task requires skill, experience, and a good eye, even with computers to help you.

One caveat before we leave the subject of paint mixing though: None of this elaborate effort will quite solve your problems if you don't know how to shoot the paint properly. Factory paint on cars is very thin and is baked on. Chances are you will be shooting your paint a lot thicker than original and will cure it at atmospheric temperatures, and these factors affect appearance. However, there are ways to counter these problems.

Tom selects a toner that will correct for the orange cast in the matching paint.

Toner is added a drop at a time and noted down so formula can be duplicated.

A perfect match in little over an hour. It used to take much longer with the old trial-and-error method.

Chapter 31
Rust
Protection

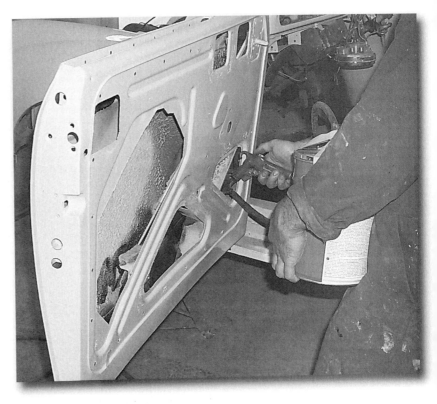

My pal Bruce Haye demonstrates the spraying of cavity wax. Note that he wears old clothes for this activity.

Rust Protection

I like to spend my U.S. winters in New Zealand where it is summer. After countless hours patching a rusted-out Morris Minor convertible I keep down there, a local friend reminded me I wasn't in sunny Southern California anymore. He cautioned me that I had better take steps to keep rust at bay if I wanted to keep New Zealand's required automotive "Warrant of Fitness" for more than a few seasons in the wet climate. I followed his recommendations. Here's what we did:

The Lilliputian machine was stripped to bare metal, new patch panels welded in, and dents taken out. Of course, stripping sheet metal leaves it naked to its worst enemy, rust. And welding causes a certain amount of oxidation too and makes sheet metal even more vulnerable. I knew that if rust wasn't stopped at this stage, my efforts would be short-lived.

Before we painted the body parts of the car we used a good rust converter to neutralize any slight surface corrosion that may have formed. Then just to be doubly sure, we went over all the panels with a wire wheel to get down into the pores of the metal and root out minute corrosion. We then shot on an etching primer to bond the paint to the metal. After that, we sprayed the car with an epoxy primer and paint that is unbelievably durable, yet flexible. But that wasn't enough. Not nearly enough for a wet climate like New Zealand's.

Cavity Wax

Since Morris Minor passenger cars—unlike the little pickups and woody Travelers—have no frames, they are particularly prone to structural problems due to corrosion because of their thin sheet metal subframe braces. The rust problems are especially serious inside box sections, cross-members and welded-in hat sections. That's because there is no way to spray paint up into them. It's a challenging situation. However, if your car is designed this way don't fret, here's the fix:

I have heard of people who drill holes and shoot in fish oil, or if short on funds, even old crankcase drippings, but in either case the smell is diabolical, and besides, the stuff will run out on a hot day and—in the case of crankcase drippings—even rinse away eventually if it gets wet. Believe it or not, motor oil is actually water-soluble. Besides, these materials are toxic and never harden, so they attract dirt, and that doesn't help matters either.

But now there is a product called cavity wax that can be shot into closed-off sections using a special wand hooked up to a source of compressed air. This amber liquid doesn't smell like a bouquet of carnations, but it isn't unbearable, and it hardens to a waxy, waterproof coating that won't melt on hot days or wash away on wet ones. The Eastwood Company sells a quality cavity wax by the way, or you can get it from

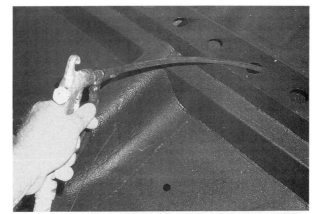

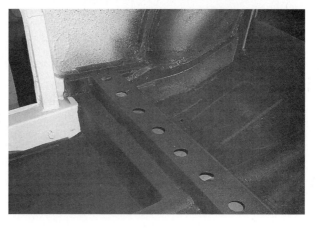

Bruce masked off the areas that didn't need underseal, then we shot it on the floor pans and structural members to prevent corrosion if my little droptop Morrie is caught out in a shower.

Take loose any access panels and liberally apply cavity wax to all surfaces to avoid future rot. This is the only access to this Morris subframe.

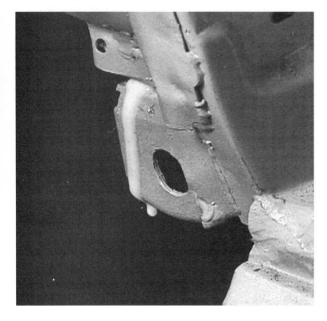

How much wax do you apply? "Shoot it on until it runs out the holes in the bottom," says Bruce.

automotive paint suppliers.

Put on a hat, eye protection and coveralls, or at least a long sleeve shirt and an old pair of work pants. You don't want to get this stuff on your skin. Cover the floor of your shop with cardboard or a cheap plastic tarp to catch the drippings, although if you splash some around, the stuff will clean up with lacquer thinner. (For tarps I just picked up some of that black plastic they sell by the yard at garden supply stores.)

Once you've put on the proper attire—using a wand especially made for the job (available from Eastwood or automotive paint suppliers) hooked up to an air compressor and a can of cavity wax— shoot the thick gooey substance into every seam, nook and cranny. Each missed cranny is an invitation to rust that you will never realize exists until it is too late. The wand comes with attachments that allow you to spray wax straight out, at right angles, and even back toward you so you can do the flip side of bulkheads through small holes in them.

Drill small holes in any box beams and cross-members on your classic so you can access these closed-off areas. Once you have thoroughly coated them inside, use rubber bungs or plugs made for the purpose to fill the holes. When you coat inside

Wax was sprayed heavily above this lap-welded seam so it would run down in and prevent oxygen and moisture from getting to the bare metal.

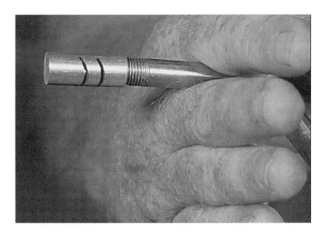

There is even a nozzle that allows you to shoot cavity wax back against a bulkhead through a small hole.

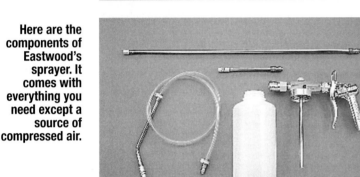

Here are the components of Eastwood's sprayer. It comes with everything you need except a source of compressed air.

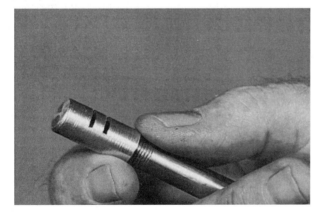

Nozzles included with the sprayer allow you to shoot straight ahead or out to the sides, as well as up and down.

doors, be sure to open the little drain holes in the bottom of the door so water that seeps in can get out. If those little drain holes are plugged you can say good-bye to your lower doors after a couple of seasons.

How do you know when you've sprayed enough wax? That's easy. When the wax runs out all over the floor you're done. I was rolling around in the stuff and it bothered me that I was paying for all that wasted wax, but I had to admit this was a very inexpensive alternative to rust removal and welding in patch panels. Be sure to coat door inner skins, inner kick panels, sills and doublers liberally. Also make sure the wax seeps well down into seams, and that all welds are truly sloshed in the stinky stuff.

Undercoating and Overcoating

Because my little Morris is a convertible, I decided it should be undercoated—or perhaps I should say overcoated, inside as well as out. That way if a shower comes along while I am paying a short visit to my favorite pub the car won't rust from the inside out.

Before shooting on the undercoating I caulked all the seams we could find with urethane sealer to make the car water- and airtight. If your car has seams that are already caulked, strip all the old caulk out before putting in fresh caulking. The stuff does eventually dry out and crumble with age. Sweep away any dust or dirt, then wash the area down with a little detergent and water. Let the seam dry thoroughly, then recaulk according to the instructions on the caulk container.

We shot on an even coat of underseal over all the floor pans and cross members on the inside. That way, even though the carpeting may get soaked and smelly if I don't air it out, all that new sheet metal underneath won't even be touched by an errant cloud burst. Of course we primed the interior surfaces with waterproof primer before shooting on the undercoating, so even if the surface of the undercoating is somehow cracked, the metal underneath will resist corrosion.

We shot on the undercoating using professional equipment, but you can do the same job using aerosol cans if you need to, though the material will be a bit thinner and more vulnerable to cracking. Make sure you mask off any areas you don't want spattered. Most modern undercoating is water-

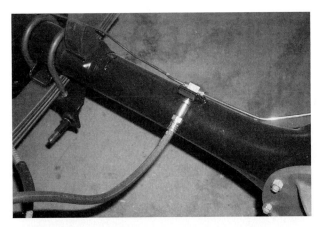

If you are doing a frame-off, don't forget to replace those brake lines too. They can burst as a result of being thinned by corrosion.

The best way to rid your car of rust is to cut it out completely, but when that isn't possible you can use Eastwood's Corroless to neutralize it.

Get into kick panels and cowl areas with cavity wax through hinge openings

My '58 Chev pickup got undercoating inside and out in the cab area to prevent rust and to deaden sound.

based, so if you are quick about it, you can clean it off of surfaces you don't want coated before it sets up using tap water. Again, make sure you coat everything liberally because any uncoated patches will amount to an Achilles heel for your chariot.

I'm just now putting the Morris back together and rubbing it out, so we haven't shot the subframe and floor pans with the underseal. We want to make sure we get all the seams covered and that there are no scraped areas. Of course, when you underseal the chassis or subframe and flooring of your classic, you will want to make sure you don't coat the engine pan, transmission, driveshaft or differential housing, because the stuff can make these items run hotter, and in the case of the driveshaft, throw it out of balance. Also, don't coat tie rod ends, shock absorbers, or any electrical items.

I undercoated my old '58 Chev pickup cab inside and out as well. And in that instance I also shot a waterproof sealant foam up in between the inner cab and the outer metal, and I filled the kick panels with it as well. The stuff seals out moisture, insulates, and cuts down on road noise too. A little extra effort in the trunk, passenger compartment and frame of a vehicle can make a big difference to its durability.

For that OEM look on gas tanks, use Eastwood's Cold Galvanizing Compound. It contains real zinc and protects the outside of the tank from rust. And for areas that are rusty where you can't get inside to cut it out, Eastwood's Corroless is a great product because it is very tough, and it converts existing rust to magnetite.

Chapter 32
Detailing & Maintenance Tips

Getting a car ready for show takes time and effort but there is no magic to it.

If you have gone to all the effort of doing a top quality paint job on a car—one that requires taking the car apart, stripping it, fixing all the imperfections and painting it properly—you will want to keep it that way. Beautiful paintwork is by far the most important aspect of any restoration because the finish work is what everyone sees. Judges at shows can be so bowled over by beautiful paint that they don't notice shortcomings in other places on a car, but if the finish work is marginal, the rest won't matter. You won't go home with a trophy.

And even if you never show your car, you will want it to maintain its beauty for years to come. Looking after a car's finish is actually an enjoyable activity for some of us, and even detailing for show—though much more intense—isn't hard if you know the techniques. To me an afternoon detailing a classic is a great stress reliever. It's my form of meditation. I get out my boom box and tune it to the golden oldies station, fill a cooler with soda, back my classic out and get to work. Soon my shoulders come down from around my ears and the troubles of the world are far behind me.

Washing

Never wash your car if you can help it. What? you ask. I repeat, never wash your car unless you have no other choice. And above all, never, ever, ever, run it through a car wash. I

have a couple of cars that were restored 15–20 years ago that have never been washed and they look absolutely stunning. They've never been washed because they have never been dirty enough to require it. I know this sounds extreme but it really isn't. Let me explain:

When you wash a car in the conventional way, which is to get out a garden hose and blast the vehicle all over, water runs down into all the little gaps and into doors and interior and cracks and crannies and causes rust, mildew and corrosion. Even rain doesn't come at a car in a steady concentrated blast, so it doesn't do as much damage as a garden hose will. The rust and corruption that develops from washing a car the conventional way is also hard to detect until it's too late.

If your car gets dusty or—heaven forbid—dirty, mix a little mild car wash solution (I like Meguiar's) in a bucket of cold water and, using a soft hand mitt, wash the dirt off using as only as much water as is necessary. Work in a small area such as one side of the hood or top, and dry it immediately with a soft Turkish towel. Don't use a chamois because they can easily drag bits of grit across a finish and scratch it.

Remove Contaminents

Next, go over the car with detailer's clay. It was invented in Japan about 20 years ago and it removes all sorts of

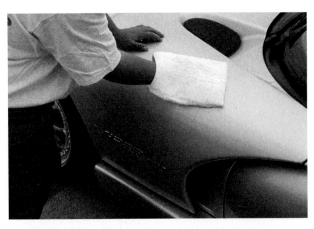

A soft mitt and as little water as possible are best for washing your car.

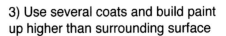
1) So you have a scratch

2) Dab in paint with soft brush

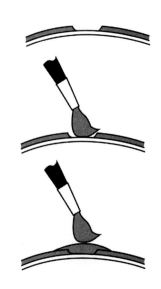
3) Use several coats and build paint up higher than surrounding surface

4) Sand bulge flush to surrounding finish with 1000 grit sandpaper

5) Finish with 2000 grit paper, then rub out and polish

Here are the steps needed to touch up stone chips. Don't neglect them because rust can develop under the finish.

environmental contaminants such as iron dust, brake pad dust, acid rain, exhaust particles and other environmental pollution that likes to embed itself in your car's paint. I first saw the stuff being used at Pebble Beach a few years ago and was amazed at how well it works.

Detailer's clay actually pulls the pollutants out of the paint which is a whole lot better than trying rub down and take paint off to get rid or them. You just give a couple of squirts of the liquid onto the finish and then rub it with a soft clay-like bar. You can feel the difference with your hand, and you can really see the difference too.

I always go over my car's finish with Detailer's clay before sealing or waxing because I would rather not seal such pollutants into the finish where they can continue to damage it. Once the car is clean, go over it with the pure carnauba wax, or Q-7 Detailer according to how the car is to be used.

Bird droppings, bugs and tar should be removed as soon as possible rather than waiting until a convenient time, because they can easily damage and alter paintwork. Keep a small container of bug and tar remover in the trunk of your car, as well as a spritzer bottle of water and some soft rags for just such emergencies. Stone chips should be fixed at the first opportunity too, so rust won't develop under the finish. See the illustration nearby for how to fix stone chips. (Note: If your car is painted with a base coat and clear you will need to use a few drops of clear over your base-coat fix as well.)

The only place a hose and a blast of water has in maintaining a car is when you need to remove dirt, salt, leaves and the like from up in fender wells, rolled fender beads and suspension. Dirt can easily build up in these areas and absorb moisture, causing rust and rot. If you get stuck in a storm or

on an unpaved road, take the wheels off your car and clean in the wheel wells before storing the car.

Waxing

Not only are show cars never washed, but they are almost never waxed either. Wax builds up and dulls a finish. Then at some point it becomes necessary to take a degreaser and wax remover to your car's finish and start over if you want it to maintain its original luster. Show cars that are driven rarely and exclusively on beautiful days only need a little Q-7 Detailer applied now and then to keep them looking beautiful. The rest of the time they are covered and kept indoors. Some restorers will use a thin coat of piano wax well rubbed in to provide added protection but that is all.

But what if you want to drive the car regularly? In that case you need the protection of a pure carnauba wax with no cleaners in it. Such a wax will help to a degree with UV protection and will act as

If you have a wax buildup on your classic you will need to strip the old wax off before detailing the car.

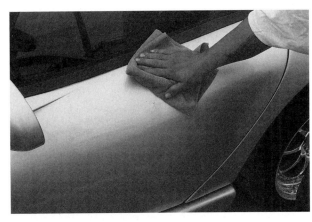

A microfiber cloth is also ideal for waxing and polishing because it will not leave fine scratches.

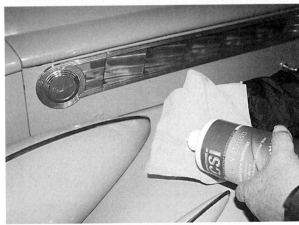

A little polish on a soft cloth will remove minor spiderwebbing and dullness.

CSI products, available through the internet, are the best there is for maintaining you car's finish.

Technique

Work in a small area and rub the wax in thoroughly. You only want a very fine layer of wax when you are finished. Wax protects as much as it's going to in very thin layers so there is no point in putting on more. If you are trying take out fine spiderweb scratches use some C21 Cut 'N Polish, then come back in with a little CSI pure wax. Apply the polish with a soft sponge, wipe it off, then come back in with a clean sponge and the wax.

If your car's finish is somewhat dull and oxidized, go over it with a buffer and foam pad using the CSI polishing compound and after that apply the wax. Never use the kind of rubbing or polishing compound on your car's fine finish that you find in the auto supply. It is far too coarse and will make more micro-scratches than it will remove.

Also, never use ordinary polishing compounds on clear coats because you can remove a lot of paint in a hurry with them, and that is what causes clear-coated cars to look as if they are molting their skin like a lizard. Once the clear coat becomes too thin to provide UV protection it starts flaking off. In fact, try not to use any polishing compound on your car if you can help it. Keep your car covered and in a garage when not in use, and keep it covered when in the hot sun too.

Beauty Is in the Details

If you are going to be showing your car, or just want it to look its very best, here are some tips from show-winning detailers that can help. How good your car looks is only dependent on how much time and effort you are willing to devote to the job of looking after it. It can easily take a week to get a car ready for a major show, and I can spend a day on routine maintenance of just one car, so don't count on being able to just wipe on and wipe off a little

a buffer between your car's finish and such things as acid rain and other contaminants. Wax can't protect your finish completely from such things, but it does help. I use Zymol when I don't have CSI, but any pure wax without cleaners will do the job.

Every good detailer has a selection of brushes adapted to cleaning wax residue out of cracks and crannies.

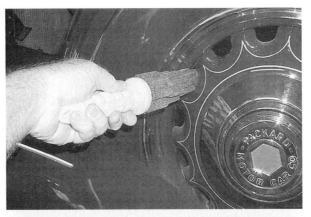

Eagle One sells its Soft-Touch wheel scrubber made of sturdy sponge for cleaning wheels.

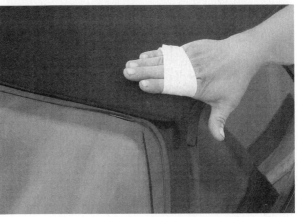

Masking tape wrapped around your hand sticky side out is great for cleaning cloth tops.

wax and have you car dazzle the judges at a show.

Clean and detail the car from the bottom up and the inside out. The final thing you should do to complete the job is tend to your car's paint. That's because you don't want to smear and smudge the finish while you are detailing the engine and chassis. Get under the car and clean and touch up the chassis and frame using some of Eastwood's Chassis Black in an aerosol can. Wipe the engine compartment down carefully, then go over the painted components with a little Armor All.

When you have the engine room sparkling, get in and brush or clean the upholstery. For leather I like Lexol's cleaners and protectants. For vinyl you can use any of the commercial concoctions available, or you can use my old tried and true Vaseline. I know that sounds awful, but when it is wiped on and rubbed out it makes a nice finish and it won't cause deterioration. Finish the interior by vacuuming and brushing the carpeting so the nap all runs the same way.

We've already gone into the particulars of waxing and glazing, but to do the job right you need a selection of brushes to get down in all the little gaps and crevices of your car's body. Every professional detailer has old toothbrushes, scrub brushes and small paintbrushes that have been cut off to make them stiffer, as well as longer ones that aren't so stiff, to get all that wax residue out of cracks and crannies.

Here are a couple of other tricks that make a big difference: Wax your car's brightwork, then come back with a little Windex to polish it to a truly bright gleam. I know that doesn't sound very sensible, but try it, you'll like the effect. Also wax your car's windows with one of those waxes with a cleaner in it. I like the Mother's brand for this. Any

tiny pits will be filled in and the glass will look brand-new.

Your car's rubber will need attention too. You can use Armor All, but I prefer saddle soap. It's a bit old-fashioned but it makes rubber look great and helps protect it. A little black silicone sealer can be pressed into cracks or gaps to make them disappear too.

Convertible tops are another set of challenges. If you have a classic show car with a natural canvas Haartz cloth top, keep it covered at all times when not in use. I've found that washing such tops almost always causes streaks and problems. If you get any dust into the fabric, vacuum it out. Use an art gum eraser like the ones you used in elementary school to erase dirt smudges, and wrap masking tape around your hand sticky-side-out to remove larger smudges, lint and dirt.

Lightly dampening a clean cloth top with a spray bottle and then parking the car in the sun can help pull out wrinkles from having the top down though. Clean bird droppings off your top immediately. They can make terrible stains in both fabric and vinyl tops. To keep vinyl tops shiny I use Vaseline. That's right. I like it better than any of the commercial products. And for plastic back windows I've found that pure carnauba wax helps keep them from fogging up.

The final piece of advice for detailing is to

remember it is called detailing for a reason. If you are going to show your car, any little missed detail will cost you points. That little bit of gasoline that dribbled down the carb when you started the car needs to be removed with a small amount of lacquer thinner. That spot of wax you missed on the bottom of a hubcap will be on top for sure when your car hits the show field. And that fine spiderwebbing in the paint will scream at you under fluorescent lighting.

When you have finished detailing the car, call your spouse or a friend and have them look over the car for anything you missed. You may well be too close to it to see certain things. Snip any stray threads sticking out of the upholstery. Be hard on yourself. You can't expect the judges to show you any mercy. And carry a kit with you to the show field that includes waxes or glazes, Windex, toothbrushes, detailer's clay, a bottle of water and a duster. I like to keep a Kozak Dry Wash cloth and a terry cloth towel on hand too, for untoward situations.

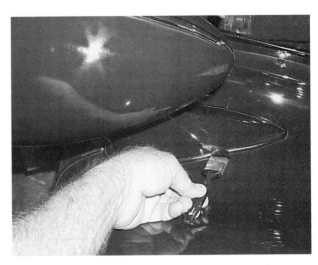

I like a cut off natural bristle paintbrush for cleaning around fender welt.

APPENDIX

THE FUTURE OF PAINT

All the new Mercedes automobiles are painted with a ceramic clear coat that is incredibly hard and scratch resistant.

There are some in the autobody trade who speculate that automotive body panels may soon be coated with thin plastic sheeting, and that is a possibility. Such technology might be cleaner and greener than what is offered today, but it is probably years away from mass production. Closer on the horizon is waterborne clear coat technology to go with the waterborne color coats. It is not required yet by any locality that I know of, but such clear coats exist, and will very likely be required in the near future for refinishing purposes for environmental reasons.

Finish, Heal Thyself

Another new technology developed by Nissan and others, and used on some of the company's upscale models as well as the new Infiniti, is self-healing paint. Toyota is apparently using it on some of their cars too.

Substances in the molecular structure of the polymers used to make the finish allow it to knit back together after it is scratched. All it generally takes is a couple of hours in the sun to make it happen. And there are two or three of these types of paints coming online for OEM use.

The molecular chemistry behind how this paint works is very complex. Suffice it to say, it works well, but only once. If you were to scratch the paint in the exact same location a second time it wouldn't work. This paint is not yet generally available at this printing, but the Japanese automakers—who developed at least one version of it—are using it, and it is being licensed in Japan and Europe for a number of uses.

Self-healing paint is not particularly hard, so it can be maintained and buffed out using the professional products recommended in the polishing chapter, but such finishes seldom need it. Toyota, Nissan and Lamborghini have been using these coatings since 2004, and they have held up well over the last few years.

A Hard Finish

In 2003, Mercedes Benz debuted a new scratch-resistant clear coat made up of sub-molecular ceramic particles that is baked on and hardened in a factory oven. The result is an extensively cross-linked, extremely durable finish that is nearly impervious to mechanical car washes and other common abuse. And at this writing, PPG has made this type of clear coat, called CeramiClear, available for domestic manufacturers as well. It is not yet in common use by refinishing shops because it requires more heat than most heat lamps are capable of producing.

Of course, ceramic clear coats aren't entirely scratchproof. They are just not as easily scratched as the

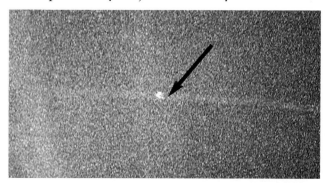

Minor scratches heal themselves with the new paints from Japan. The top photo shows a minor scratch, the bottom photo shows the same spot after just a few hours in direct sunlight.

urethanes in common use by refinishers today. And fortunately for people who need to polish out their SL coupe, CSI Clearcoat Solutions makes a product specially designed for such finishes that will polish them quickly and effectively. They call it their Ceram-X Polish 62-203 and it is available directly from the manufacturer (see page 148) or from professional jobbers. However you won't find it at Pep Boys.

The professional always has the benefit of the educational resources of the major automotive paint manufacturers, and sometimes amateurs are welcome at their seminars as well. If there is a new paint and finish technology available that you want to know more about, look up the manufacturer on the Internet and ask them to send as much print information as possible. And then get to know some pros. They will often be able to get you into teaching sessions and seminars on new techniques and finishes. Automotive finishes have changed dramatically and extensively in the last twenty years, and the trend is accelerating. It pays to keep learning.

CSI Ceram X and the correct buffing pad are all you need to polish the new super-hard ceramic coatings being used on Mercedes and Audis.

RESOURCES

3M Customer Service
1-888-3M-HELPS
(1-888-364-3577)
www.3m.com

Back-O-Tools
P.O. Box 501 Dept 8
Turlock CA 95381
1-209-632-0120

California Car Cover Co.
9525 De Soto Avenue
Chatsworth CA 91311
1-800-423-5525
1-818-998-2100
(Sells excellent car covers for all makes and years.)

Car Care Products
13 Harbor Park Drive
Port Washington NY 11050
1-888-484-9560
1-516-484-9500
www.autobarn.com
(Good source for car-care products.)

Classic Motoring Accessories LTD.
5008 West Linebaugh Avenue
Suite 60
Tampa FL 33624
1-800-628-7596
www.properautocare.com
(Good source of car-care products in the Southeast.)

CSI Clearcoat Solutions Inc.
Los Angeles CA 90023
www.clearcoatsolutions.com
1-877-274-4296
(Best one-step professional polishing compound, wax and detailer available.)

DeVilbiss Spray Guns
ITW Automotive Refinishing
1724 Indianwood Circle
Maumee OH 43537-4005
Customer Service: 1-800-992-4657
Tech Support: 1-888-992-4657

The Eastwood Company
263 Shoemaker Road
Pottstown PA 19464
1-800-343-9353
1-610-323-2200
www.eastwood.com
(Best source of automotive restoration and refinishing tools around.)

Evapo-rust
Orison Marketing
4801 South Danville Drive
Abilene TX 79602
1-800-460-2403
www.orisonllc.com
(Sells top quality, environmentally safe rust removers.)

Hawkeye Borescopes
Gradient Lens Corporation
207 Tremont Street
Rochester NY 14608
1-800-536-0790
www.hawkeyblue.com
(Source for a great tool to peer inside areas of car bodies you can't normally see.)

Lincoln Welders
Help Line: 1-800-833-9353
1-800-833-9353
(Makers of the Mig Pak 10 and Pak 15 welders.)

PPG Automotive Paints
1 PPG Place
Pittsburgh PA 15272
1-412-434-3131
www.ppg.com
(Major supplier of automotive paints.)

Sherwin-Williams Automotive Finishes
Attn. Customer Service
4440 Warrensville Center Road
Warrensville Heights OH 44128-2837
1-800-798-5872 (Western)
1-800-526-5493
E-mail: autocontact@sherwinwilliams.com

Sikkens Car Refinishes
MIXIT 1-800-866-4948
1-800-618-1010
www.sikkens.com
(Good source for high quality automotive
paints.)

Steele Rubber Products
6180 East NC 150 Hwy
Denver NC 28037
1-800-544-8665
1-704-483-9343
www.steelerubber.com
(A good supplier of rubber weather-stripping
and seals for most cars ever made.)

Survivair
3001 South Susan Street
Santa Ana CA 92704
Customer Service: 1-800-821-7236
Tech Support: 1-800-375-6020
(Makers of the finest paint respirators and
positive pressure painting systems.)

Williams Lowbuck Tools Inc
4175 California Avenue
Norco CA 92860
1-951-735-7848
(Source for double-cut metal shears.)

METRIC CUSTOMARY UNIT EQUIVALENTS

Multiply:	by:	to get:	Multiply by:	to get:
LINEAR				
inches	x 25.4	= millimeters (mm)	x 0.03937	= inches
feet	x 0.3048	= meters (m)	x 3.281	= feet
yards	x 0.9144	= meters (m)	x 1.0936	= yards
AREA				
inches^2	x 645.16	= millimeters^2 (mm^2)	x 0.00155	= inches^2
feet^2	x 0.0929	= meters^2 (m^2)	x 10.764	= feet^2
VOLUME				
inch^3	x 0.01639	= liters (l)	x 61.023	= liters
feet^3	x 28.317	= liters (l)	x 0.03531	= feet^3
feet^3	x 0.02832	= meters^3 (m^3)	x 35.315	= feet^3
fluid oz	x 29.57	= milliliters (ml)	x 0.03381	= fluid oz
quarts	x 0.94635	= liters (l)	x 1.0567	= quarts
gallons	x 3.7854	= liters (l)	x 0.2642	= gallons
MASS				
ounces (av)	x 28.35	= grams (g)	x 0.03527	= ounces (av)
pounds (av)	x 0.4536	= kilograms (kg)	x 2.2046	= pounds (av)
FORCE				
ounces-f (av)	x 0.278	= newtons (N)	x 3.597	= ounces-f (av)
pounds-f (av)	x 4.448	= newtons (N)	x 0.2248	= pounds-f (av)

TEMPERATURE
Degrees Celsius (C) = 0.556 (F − 32) Degree Fahrenheit (F) = (1.8C) + 32

ENERGY OR WORK (Watt-second = joule= newton-meter)				
foot-pounds	x 1.3558	= joules (J)	x 0.7376	= foot-pounds
Btu	x 1055	= joules (J)	x 0.000948	= Btu
PRESSURE OR STRESS				
pounds/sq in.	x 6.895	= kilopascals (kPa)	x 0.145	= pounds/sq in
TORQUE				
pound-inches	x 0.11298	= newton-meters (N-m)	x 8.851	= pound-inches
pound-feet	x 1.3558	= newton-meters (N-m)	x 0.7376	= pound-feet
pound-inches	x 0.0115	= kilogram-meters (Kg-M)	x 87	= pound-feet
pound-feet	x 0.138	= kilogram-meters (Kg-M)	x 7.25	= pound-feet
POWER				
horsepower	x 0.74570	= kilowatts (kW)	x 1.34102	= horsepower

COMMON METRIC PREFIXES
mega (M)   = 1,000,000 or 106 centi (c) = 0.01 or 10-2
kilo (k)   = 1,000 or 103 milli (m) = 0.001 or 10-3
hecto (h) = 100 or 102 micro (u) = (u)0.000,001 or 10-6

ABOUT THE AUTHOR

Jim Richardson is the author of many books on auto restoration, including *Classic Car Restorer's Handbook*, *Pro Paint & Body* and the *Classic Chevy Truck Handbook* (the last two are still available from HPBooks). He is also a frequent contributor to such magazines as *Hemmings Classic Car* and *Auto Restorer*. He has appeared on TV shows such as *My Classic Car* and *Classic Car Restorer*. Richardson divides his time between his homes in Southern California and New Zealand and still enjoys the car culture in which he grew up.

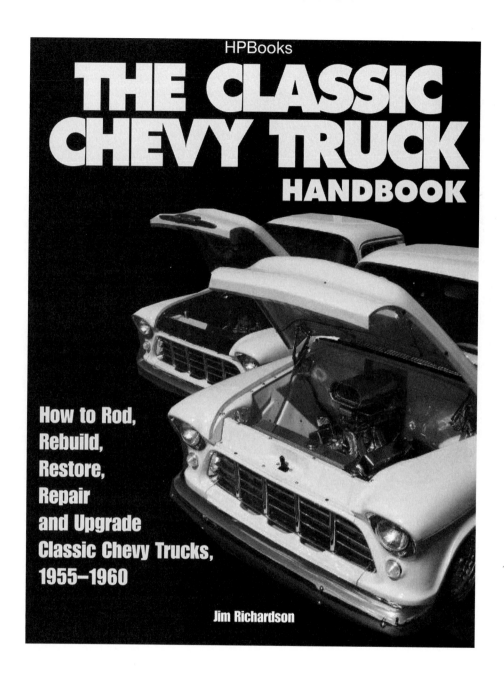

GENERAL MOTORS

Big-Block Chevy Engine Buildups: 978-1-55788-484-8/HP1484
Big-Block Chevy Performance: 978-1-55788-216-5/HP1216
Building the Chevy LS Engine: 978-1-55788-559-3/HP1559
Camaro Performance Handbook: 978-1-55788-057-4/HP1057
Camaro Restoration Handbook ('61–'81): 978-0-89586-375-1/HP758
Chevy LS Engine Buildups: 978-1-55788-567-8/HP1567
Chevy LS Engine Conversion Handbook: 978-1-55788-566-1/HP1566
Chevy LS1/LS6 Performance: 978-1-55788-407-7/HP1407
Classic Camaro Restoration, Repair & Upgrades:
  978-1-55788-564-7/HP1564
The Classic Chevy Truck Handbook: 978-1-55788-534-0/HP1534
How to Rebuild Big-Block Chevy Engines:
  978-0-89586-175-7/HP755
How to Rebuild Big-Block Chevy Engines, 1991–2000:
  978-1-55788-550-0/HP1550
How to Rebuild Small-Block Chevy LT-1/LT-4 Engines:
  978-1-55788-393-3/HP1393
How to Rebuild Your Small-Block Chevy:
  978-1-55788-029-1/HP1029
Powerglide Transmission Handbook: 978-1-55788-355-1/HP1355
Small-Block Chevy Engine Buildups: 978-1-55788-400-8/HP1400
Turbo Hydra-Matic 350 Handbook: 978-0-89586-051-4/HP511

FORD

Classic Mustang Restoration, Repair & Upgrades:
  978-1-55788-537-1/HP1537
Ford Engine Buildups: 978-1-55788-531-9/HP1531
Ford Windsor Small-Block Performance:
  978-1-55788-558-6/HP1558
How to Build Small-Block Ford Racing Engines:
  978-1-55788-536-2/HP1536
How to Rebuild Big-Block Ford Engines:
  978-0-89586-070-5/HP708
How to Rebuild Ford V-8 Engines: 978-0-89586-036-1/HP36
How to Rebuild Small-Block Ford Engines:
  978-0-912656-89-2/HP89
Mustang Restoration Handbook: 978-0-89586-402-4/HP029

MOPAR

Big-Block Mopar Performance: 978-1-55788-302-5/HP1302
How to Hot Rod Small-Block Mopar Engine, Revised:
  978-1-55788-405-3/HP1405
How to Modify Your Jeep Chassis and Suspension For Off-Road:
  978-1-55788-424-4/HP1424
How to Modify Your Mopar Magnum V8:
  978-1-55788-473-2/HP1473
How to Rebuild and Modify Chrysler 426 Hemi Engines:
  978-1-55788-525-8/HP1525
How to Rebuild Big-Block Mopar Engines:
  978-1-55788-190-8/HP1190
How to Rebuild Small-Block Mopar Engines:
  978-0-89586-128-5/HP83
How to Rebuild Your Mopar Magnum V8:
  978-1-55788-431-5/HP1431
The Mopar Six-Pack Engine Handbook:
  978-1-55788-528-9/HP1528
Torqueflite A-727 Transmission Handbook:
  978-1-55788-399-5/HP1399

IMPORTS

Baja Bugs & Buggies: 978-0-89586-186-3/HP60
Honda/Acura Engine Performance: 978-1-55788-384-1/HP1384
How to Build Performance Nissan Sport Compacts, 1991–2006:
  978-1-55788-541-8/HP1541

How to Hot Rod VW Engines: 978-0-91265-603-8/HP034
How to Rebuild Your VW Air-Cooled Engine:
  978-0-89586-225-9/HP1225
Porsche 911 Performance: 978-1-55788-489-3/HP1489
Street Rotary: 978-1-55788-549-4/HP1549
Toyota MR2 Performance: 978-155788-553-1/HP1553
Xtreme Honda B-Series Engines: 978-1-55788-552-4/HP1552

HANDBOOKS

Auto Electrical Handbook: 978-0-89586-238-9/HP387
Auto Math Handbook: 978-1-55788-020-8/HP1020
Auto Upholstery & Interiors: 978-1-55788-265-3/HP1265
Custom Auto Wiring & Electrical: 978-1-55788-545-6/HP1545
Electric Vehicle Conversion Handbook: 978-1-55788-568-5/HP1568
Engine Builder's Handbook: 978-1-55788-245-5/HP1245
Fiberglass & Other Composite Materials: 978-1-55788-498-5/HP1498
High Performance Fasteners & Plumbing: 978-1-55788-523-4/HP1523
Metal Fabricator's Handbook: 978-0-89586-870-1/HP709
Paint & Body Handbook: 978-1-55788-082-6/HP1082
Practical Auto & Truck Restoration: 978-155788-547-0/HP1547
Pro Paint & Body: 978-1-55788-394-0/HP1394
Sheet Metal Handbook: 978-0-89586-757-5/HP575
Welder's Handbook, Revised: 978-1-55788-513-5

INDUCTION

Engine Airflow, 978-155788-537-1/HP1537
Holley 4150 & 4160 Carburetor Handbook: 978-0-89586-047-7/HP473
Holley Carbs, Manifolds & F.I.: 978-1-55788-052-9/HP1052
Rebuild & Powertune Carter/Edelbrock Carburetors:
  978-155788-555-5/HP1555
Rochester Carburetors: 978-0-89586-301-0/HP014
Performance Fuel Injection Systems: 978-1-55788-557-9/HP1557
Turbochargers: 978-0-89586-135-1/HP49
Street Turbocharging: 978-1-55788-488-6/HP1488
Weber Carburetors: 978-0-89589-377-5/HP774

RACING & CHASSIS

Advanced Race Car Chassis Technology: 978-1-55788-562-3/HP562
Chassis Engineering: 978-1-55788-055-0/HP1055
4Wheel & Off-Road's Chassis & Suspension: 978-1-55788-406-0/HP1406
How to Make Your Car Handle: 978-1-91265-646-5/HP46
How to Build a Winning Drag Race Chassis & Suspension:
The Race Car Chassis: 978-1-55788-540-1/HP1540
The Racing Engine Builder's Handbook: 978-1-55788-492-3/HP1492

STREET RODS

Street Rodder magazine's Chassis & Suspension Handbook: 978-1-55788-346-9/HP1346
Street Rodder's Handbook, Revised: 978-1-55788-409-1/HP1409